# Surfaces
# and their
# Measurement

KOGAN PAGE SCIENCE PAPER EDITION

# Surfaces and their Measurement

### David Whitehouse

First published in 2002 by Hermes Penton Ltd
Paperback edition first published in 2004 by Kogan Page Science, an imprint of Kogan Page Ltd

Kogan Page Science
120 Pentonville Road
London N1 9JN
United Kingdom
www.koganpagescience.com

Kogan Page US
22883 Quicksilver Drive
Sterling VA 20166-2012
USA

© Taylor Hobson Ltd, 2002

Colour plates provided by Taylor Hobson Ltd

**British Library Cataloguing in Publication Data**

A CIP record for this book is available from the British Library.

ISBN 1 9039 9660 0

Typeset by Saxon Graphics Ltd, Derby
Transferred to digital printing 2006
Printed and bound by CPI Antony Rowe, Eastbourne

# Contents

# Foreword

The importance of surface metrology has been acknowledged in the UK for many years. A lot of the action has been centred on a few firms such as Taylor Hobson, Hilger Watts, Mercer etc. Also, there was an early awareness by non-commercial agencies including, of course, the National Physical Laboratory (NPL). However, what is not so well known is the priority given to surface metrology. (Table 0.1 gives some key dates.)

In January 1939, the Research Department of the Institution of Production Engineers was set up under the direction of Dr. Georg Schlesinger. Surface finish was given top priority. By March 1940, the first formal report on surface metrology was produced and published [0.1]. This report set out the following aims:

(a) to replace the loose descriptive methods of defining surface finish by a more definite system,
(b) to select suitable units for the measurement of surface finish,
(c) to suggest appropriate symbols for use on drawings,
(d) to compare various methods for observing the features of surface structure.

The main problems were seen as:

(e) the provision of standards for the measurement of surface finish,
(f) the determination of the type of finish best suited to any given application.

In those days, over 90% of the fields of interest in using and classifying surface finish were purely mechanical and the scales of size in the ten thousandths of inches.

Nowadays, there has been an expansion of relevance to include other fields such as semiconductors, electronics and optics. Also, the scales of size are now down to nanometric and even atomic levels. Yet it seems that we still have to address the same problems, given perhaps some more sophisticated tools.

This book is an attempt to progress some of the original aims with today's problems.

The book has been written in a relatively large number of small chapters. This is intentional so that each chapter can be considered as a stand-alone contribution to the subject. There are, however, distinct groupings that reflect the way in which the subject evolved and the way in which it is now being used.

Two such groupings are surface finish and roundness; these being the two base geometries. Other chapters can be related to the principal chapters under the base heading e.g. cylindricity with roundness.

Different types of grouping are possible with respect to, for example, instrumentation i.e. stylus methods and optical methods. Because these subjects intertwine with the geometrical features, it has been decided to enable connections to be made by using small chapters rather than large all-embracing ones.

The book is primarily intended to help designers and inspectors to understand the fundamentals of surface metrology and, where appropriate, point to relevant procedures for specification. The mathematics has been kept to a minimum in order not to disrupt the flow of information. The references given and the advice in the text should be sufficient to allow the reader to follow the subject with greater rigour, if deemed necessary.

The fact that the book is primarily concerned with helping the reader to get an understanding of the subject, rather than giving sets of rules, makes it suitable for use in degree courses in manufacturing engineering and mechanical engineering.

*David Whitehouse*

## References

0.1  Schlesinger G. *Surface Finish Journal of Inst. of Prod. Eng.* xix No. 3 p99-106.

## Acknowledgements

Where appropriate, sources of figures and text are acknowledged in situ. Colour plates are provided by Taylor Hobson Ltd, with thanks.

**Table 0.1** *Some early dates of importance in the metrology and production of surfaces*

| | |
|---|---|
| 1731 | Introduction of Vernier system of linear measurement by Pierre Venier |
| 1769 | Smeaton's first boring machine for cannon |
| 1775 | Watts steam engine based on Wilkinson machine |
| 1800 | High-carbon steels used for cutting tools |
| 1865 | Robert Musket of Sheffield introduced semi-high-speed steel |
| 1867 | Vernier calipers manufactured by Brown and Sharp |
| 1886 | Reynolds' paper to Royal Society on hydrodynamic theory |
| 1890 | Introduction of synthetic abrasive grinding materials |
| 1895 | Micrometer introduced |
| 1896 | Introduction of gauge blocks by Johanson – progressive tolerance principle |
| 1898 | Chip analysis mode of turning |
| 1900 | Introduction of high-speed tool steel |
| 1904 | Nicolson analysis of tool pressures in chip removal |
| 1911 | Commercial manufacture of gauge blocks |
| 1915 | Commercial introduction of centreless grinding |
| 1916 | Development of cemented carbides for cutting tools in Germany |
| 1922 | Introduction of first lapping machine |
| 1929 | First tracer-type surface machine by Schmalz in Germany |
| 1933 | Abbott's profilometer conceived in USA |
| 1934 | Linnik surface microscope in USSR |
| 1934 | Gloss method, Carl Zeiss, Germany |
| 1935 | Flemming's tracer profilometer, Germany |
| 1936 | Brush surface analyser in USA |
| 1936 | Perthen capacitance surface gauge, Germany |
| 1936 | Superfinishing process introduced by Chrysler Corporation in USA |
| 1938 | Tomlinson surface finish recorder in UK |
| 1939 | Nicolau pneumatic gauge, France |
| 1940 | Talysurf developed in UK |
| 1940 | First B46 standard on surface finish, USA |
| 1942 | Machined surface standards, Norton, USA |
| 1943 | Use of diamond cutting tools reported, UK |
| 1944 | First replication methods for surfaces 'FAX FILM' |

# 1

---

# Introduction

## 1.1 General

The object of this book is to clarify the situation regarding surface metrology manufacture and performance. It is not to provide intricate detail on procedures or instruments but more to explore the reasons behind specifications and standards. When and where to apply 'one number' values will be indicated. Necessary discrimination between parameters will be described, wherever appropriate, and redundant discrimination highlighted.

In particular, the whole philosophy of the book will be to separate out the uses of surface measurement. Manufacturing aspects will be considered apart from functional effects. The reasons for the dichotomy of use and the problems caused by failure to recognize the difference in use will be discussed.

Surface parameters have been used as attributes and not as variables; the trouble is that when to use the one or the other has never been clear and is still not clear.

## 1.2 What is surface metrology?

The expression 'surface metrology' usually conjures up pictures of rough surfaces probably having regular triangular or sine wave shapes. These simple pictures correspond to the mark produced by the tool in removing material. In addition, longer waves are produced by errors in the path of the tool, which in effect cause departures from its intended shape. These could be roundness errors, straightness errors or even errors of form. Figure 1.1 shows a general shape, highly magnified.

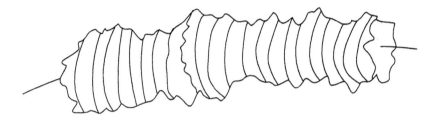

**Figure 1.1** *Surface errors in cylinder*

This is very far from an ideal cylinder.

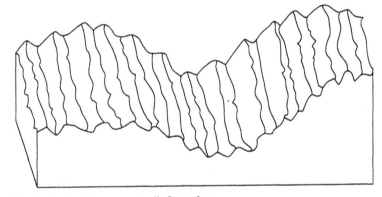

**Figure 1.2** *Surface deviations on nominally flat surface*

Figure 1.2 shows exactly the same sort of deviations on a nominally flat surface.

Many sorts of geometric feature can be considered to be surface deviations. Examples are lack of squareness, errors in taper or conicity, parallelism and roundness error – in fact a great many deviations that do not include size. This characterization is not so straightforward as it looks at first sight. For example, the radius of a part can be considered to belong to dimensional metrology whereas local curvature of the part comes under surface metrology.

## 1.3   Usefulness of surfaces

The importance of surface roughness is often neglected. It has been regarded as being on the fringe of general engineering, being looked on as an irritant that has to be dealt with but with not too much trouble.

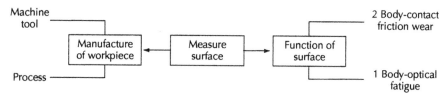

**Figure 1.3**  *Surface measurements in manufacture and performance sequence*

'Make the finish as fine as possible – it is bound to be better.' This statement has been made often because of ignorance: not intentionally but used nevertheless. Factors that could have shown that the surface was important simply were not in place. These include digital analysis and tribological knowledge – random process analysis. It is only within the past ten years that the tools for surface investigation have been available. The result is shown in a block diagram (Figure 1.3). This shows surface measurement in between the manufacture of a part and its performance (here and from now on called the **function**).

It is now reasonably established that the surface metrology has two roles; one to help control the manufacture, including the process and the machine tool, and the other to help optimize the function. These two roles can have a profound impact on quality. Figure 1.4 shows how surface metrology does this. Controlling the manufacture helps repeatability and hence quality of conformance. Functional optimization helps the designer and thereby assists in the quality of design. The former might be obvious but the latter is not. It is only very recently that the significance of this predictive role of metrology has emerged [1.1]. Another aspect of the influence of the surface texture is that sometimes it helps function, sometimes it will be detrimental. Notice how by looking at two different aspects of the surface data, namely the spread of readings and their value, the two roles can be separated. At this stage it does not matter which parameter of the surface is being measured.

In case this explanation seems woolly, consider the example of turning. If the cutting is smooth, with no built-up edge or chipped tools, the surface will be regular and more or less constant over an area. Therefore, the surface parameters will show little spread. The small spread therefore indicates a process under control.

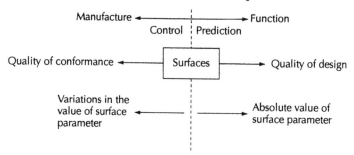

**Figure 1.4**  *Roles of surface metrology*

The relationships between manufacture measurement and function is shown in the three figures comprising Figure 1.5. In Figure 1.5(a), the conventional situation is shown. This is basically a trial and error approach usually augmented by experience. Figure 1.5(b) demonstrates the use of metrology to encompass the machine tool as well as the process. Figure 1.5(c) brings in the function. Ultimately, the emphasis will shift from Figure 1.5(a) to 1.5(c) because it is the function that is important rather than the manufacture. In what follows in 1.3.1, the process will be investigated.

In the case of function, imagine a bearing in which an oil film, between the rotating shaft and the journal, is generated by the movement. If, say, the peak to valley height of the roughness on the shaft is bigger than the thickness of the oil film then seizure will occur. There will be metal to metal contact and poor performance. Arranging that the roughness is much smaller than the anticipated oil film will ensure that the bearings work well. Here it is the actual value of the roughness that is important i.e. it has to be smaller than the film thickness.

(1) Past and present situation

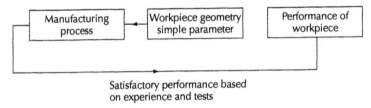

Figure 1.5(a)   *Manufacture measurement and function: conventional situation*

(2) Future situation – manufacturing control

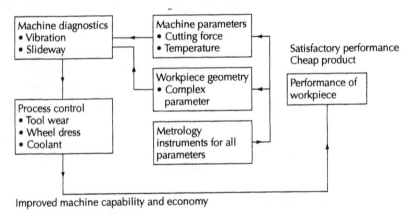

Figure 1.5(b)   *Use of metrology to encompass machine tool in process*

## (3) Function optimization

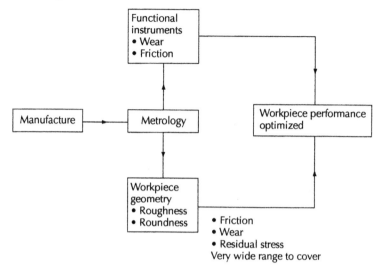

**Figure 1.5(c)**   *The introduction of function into process*

### *1.3.1  Process*

In the figures, it is straightforward to relate the roughness measurement to the process marks. Turning, for example, can be approximate to a triangular profile:

**Figure 1.6**   *Triangular approximation to profile*

A swift calculation reveals that $R_a$, the average roughness, value is simply A/2. A similar calculation can be carried out for a sine wave profile as shown:

**Figure 1.7**   *Sine wave profile*

Here, the $R_a$ value is $\dfrac{2A}{\pi}$. It is straightforward to visualize the tool making these profile marks. The feed determines the value $A$, and can be found by observation. It could be argued that quality of conformance for the process could be controlled just by looking at the profile. Unfortunately, subjective judgements cannot be included effectively in a quality specification, hence the need for a number like $R_a$. In earlier days, the spacing was not considered to be important.

### 1.3.2  Machine tool

The role of surface parameters in machine tool monitoring is not so easy. The profile by itself is not adequate, as will be seen later. Ideally, the instrument stylus should be able to follow the intended path of the tool. Any deviation from that specified on the drawing could be attributed to a problem. Unfortunately, this does not happen in practice. One or more profiles are taken, usually at fixed spacing across the surface as shown in Figure 1.8(a).

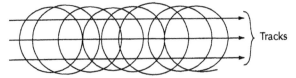

**Figure 1.8(a)**  *Profiles taken at fixed spacing*

Here, a milled surface is shown. There is no attempt to follow the clear tool path. The result is a real mess; process marks and tool path marks are jumbled together and somehow these should be separated. Needless to say, they are not. There are pattern recognition methods based on optical diffraction, for example, that could distinguish these two effects but they are not ideal. It is not the tool mark pattern that is wanted. It is the deviations from it! Areal (or 3D) mapping helps to visualize the 'lay' pattern but hardly enough to justify the effort. A good magnifying glass of ×10 is more than adequate if it is only the visual impact that is required.

Often the machine tool path is at right angles to the process mark as in turning, for example. Under these circumstances, it is possible to make some distinction. A good example, which will be discussed later, is the use of the roundness value for the machine tool path and the roughness for the process.

### 1.3.3 Function

Functionality of surfaces is strewn with problems. It could justifiably be asked what surface metrology has got to do with the optimization of performance (function).

Ideally, the way to test for functionality e.g. resistance to wear is to try it! In other words, the test should mimic the function in terms of loads, speeds and materials. Then the test should be carried out to see if it works using a wear machine to test for wear, a friction machine to test for friction and so on. Unfortunately, except in a few areas, this is not possible, the spread of parameter values and rig configurations is too large for any instrument maker to contemplate this approach as a commercial enterprise.

Instead of the direct approach, it is possible to use surface metrology. The indirect approach has been substituted for the direct. This alternative is to measure the surface and then, by virtue of some theory or experience, estimate the likely performance of the surface. Hence the growth of surface measuring instruments. Unfortunately, the theories and the experience are scarce so that often parameters are wrongly ascribed to a particular function.

Worse still is when the process control parameters such as $R_a$ are given functional roles that they were never intended for. Hence the confusion in the minds of designers when specifying surface finish.

What will probably happen is that instruments will be used to map the surface and the raw data will be used in a simulated functional test in the computer.

## 1.4    Nature of surfaces

### 1.4.1    General

Before considering what surfaces are, it is useful to determine first what surfaces are not.

The problem is concerned with visualization. Most engineers and people such as metallurgists are familiar with the appearance of a machined surface, whether just by eye or through a microscope. These pictures give a reasonable idea of spacings on the surface as well as the pattern of the lay but not much idea of the 'roughness', which is usually perceived as the heights of the machining marks. But the height is what the engineer needs for tolerancing, assembly, or simply to check on the process.

This shortcoming was easily overcome by surface instrument makers by providing a magnified chart of the height deviations. Unfortunately, as the heights are small compared with the spacings – about 100:1 – the horizontal magnification that is needed is much smaller than that needed to see the vertical heights clearly. As the objective of the exercise was to magnify the heights so that the deviations were clearly visible on the chart, what was the criterion for the horizontal magnification? This was the length of chart that could be placed on a worktable and viewed from end to end without moving the head. In other words, the whole profile picture could be viewed in its entirety at a glance. This enabled visual correlations to be judged, patterns to be seen and detail to be assimilated without losing the spatial reference of the stationary head. An angle of about 120° is regarded as suitable for the viewing angle (Figure 1.8(b) overleaf), which meant that the magnified chart at the viewing magnification is about one yard or metre long.

The pragmatic approach determined the chart length, which then obviously fixed the horizontal magnification. So, the resultant chart had a vertical magnification much greater than the horizontal magnification ~ 100:1, which meant that the profile of the surface is completely distorted. The engineer found this distortion useful. It highlighted the peaks: the spacings were largely ignored.

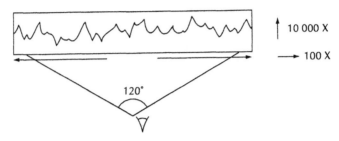

**Figure 1.8(b)**   *Viewing angle of chart*

The troublesome by-product of this distortion is that many engineers pictured surfaces as having very sharp peaks as in the Alps rather than in reality being more like the hills of the Peak District in Derbyshire (Figure 1.9).

True shape                                                    Distorted shape

**Figure 1.9**   *True and distorted surfaces*

Peaks that could be knocked off, crevice-like valleys and sharp slopes were all identified as normal, even by tribologists. The onset of plastic deformation in contact theory was taken to be almost instantaneous, whereas in fact elastic deformation was much more likely to occur if the true shape of the surface was considered.

| Process | Roughness ($R_a$) in µm | | | | | | | | | |
|---|---|---|---|---|---|---|---|---|---|---|
| | 0.05 | 0.1 | 0.2 | 0.4 | 0.8 | 1.6 | 3.3 | 6.3 | 12.5 | 25 |
| Superfinishing | | | | | | | | | | |
| Lapping | | | | | | | | | | |
| Polishing | | | | | | | | | | |
| Honing | | | | | | | | | | |
| Grinding | | | | | | | | | | |
| Boring | | | | | | | | | | |
| Turning | | | | | | | | | | |
| Drilling | | | | | | | | | | |
| Extruding | | | | | | | | | | |
| Drawing | | | | | | | | | | |
| Milling | | | | | | | | | | |
| Shaping | | | | | | | | | | |
| Planing | | | | | | | | | | |

**Figure 1.10**   *Typical roughness values obtainable by different finishing processes*

Figure 1.10 shows a list of surface roughnesses of different processes. It can be seen at a glance that there is a range of heights from 0. 05μm to 25μm i.e. 500:1. When the newer variants of these processes such as diamond turning or ductile grinding are taken into account along with atomic type processes such as energy beam milling, the range becomes much greater. It is more likely that 50000:1 is the current range.

Such a range imposes a lot of problems on the instrument maker. To abide by well-established rules of metrology over all these surfaces is virtually impossible. For example, the unit of metrology should be of the same order of magnitude as the feature being measured. This would be difficult over such a range. It is no wonder that stylus, optical, X-rays and ultrasonics all have to be used.

The real problem is not so much size as it is shape. Modern multiprocesses such as plateau honing and other 'designer surfaces' have asymmetrical characteristics, unlike grinding, which is largely symmetrical (Figure 1.11).

**Figure 1.11(a)**   *Conventional grinding*

**Figure 1.11(b)**   *Plateau honing*

Ways of specifying these characteristics will be given later on but even more trouble-some is the use of composite or porous materials, which can have 'hidden' features.

Figure 1.11(c) shows a cross section inside the skin of the surface of, say, cast iron. Re-entrant features are present but not measurable with conventional stylus or optical methods.

**Figure 1.11(c)**   *Cross section inside surface skin*

Remember that it was straightforward to infer the tool and process parameters from the surface geometry in turning. It is impossible from the surface above. It seems that short of a breakthrough in ultrasonics or X-rays, the only way to estimate the surface geometry above is as the process is taking place – in much the same way as rapid prototyping 'sees' inside designs as they are synthesized. Luckily, most applications involve contact where such surfaces as in Figure 1.11(c) are not used. But, in lubrication and plating it is quite a different story!

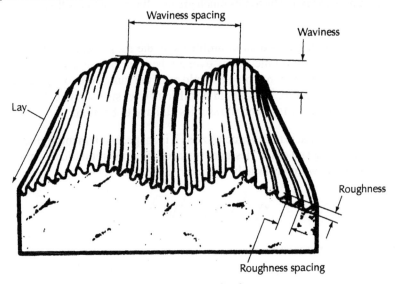

Figure 1.12(a)   *Roughness and waviness, the constituents of surface texture*

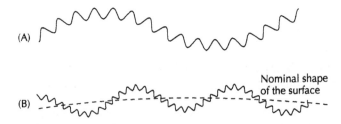

Figure 1.12(b)   *Surface components*

Many surfaces exhibit both roughness and waviness (A) and some also have a form error (B).

The surface cannot be separated from the manufacture. Process and machine tool/effects are always present. The former is called the **roughness** and the latter the **waviness** (Figure 1.12(a)). The roughness is inevitable – it is the mark of the process but the waviness – usually of longer wavelength – is a result of a problem with the machine tool and in principle could be avoided. It is called waviness because it is often periodic in nature, usually a vibration caused by lack of stiffness or a balance problem. It can be a slowly moving random type wave, which looks 'wavy', caused by deflections produced by asymmetric forces. Waviness is, in effect, errors in the path of the surface generator i.e. the tool path in turning or milling and even though it is often at right angles to the process mark, a component of it shows through when measuring across the 'lay'. One example is shown schematically in Figure 1.12(b).

The longest wavelengths involved in waviness deflect elastically under load rather than crushing as can be the case with roughness marks.

Also, in addition to roughness and waviness, even longer wavelengths can be introduced into the surface geometry by weight deflection or long-term thermal effects. These cause errors in the general shape of the part, that is, deviations from the shape required and specified by the designer. Figure 1.13 shows a typical breakdown of a ground surface.

The nature of the complete surface is usually complicated and is difficult to assess as is shown later. Figure 1.14 is a picture of a surface produced by multiple tracking with a stylus or optical measuring instrument.

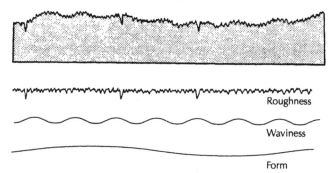

Roughness

Waviness

Form

**Figure 1.13** *Typical breakdown of a surface*

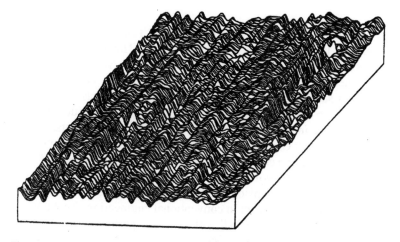

**Figure 1.14** *Parallel traces across a ground surface*

## 1.4.2 Scales of size, trends due to miniaturization

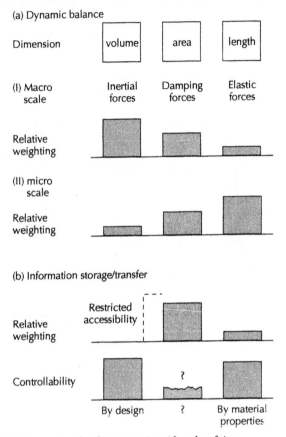

**Figure 1.15**  *The importance of surface properties with scales of size*

The nature of surfaces is changing, even in engineering. They are becoming more important rather than less. This is due to miniaturization. The dynamic nature of forces changes as a function of the physical size of the object. At typical engineering sizes of centimetres, inertia is most important, followed by damping terms, then elastic terms as seen in Figure 1.15(a). So, for normal objects, it is inertial forces related to volume that dominate movement. Taking the scale down below 1mm produces a change in the balance of the three force components. Elastic effects become the dominant component. Heavy mass effects and spring modules are easy to understand by design engineers. What are less understood are the forces associated with area rather than volume (for mass and inertia) or length (elasticity). Such area forces are surface tension and friction-producing effects like wear. These are difficult enough to deal with in conventional engineering situations but are even more difficult to control at small scales. This is partially a result of the surface finish, which dominates areal effects. So, as the scale of the workpiece decreases, more attention should be placed on the surface finish, not less. Unfortunately, whereas it is easy to reduce the size of a part, it is not easy to reduce the texture size by the same proportion; it is not just the surface geometry that is important, it is its relationship to the dimension of the part.

So, while inertial and elastic effects can be controlled by the material properties, there is no such control over surface effects and yet in most applications transfer of forces, energies and information take place at the interface between parts where the surfaces make contact.

**Table 1.1**   *Functional metrologies*

| Property | Composition | Cause | Result |
|---|---|---|---|
| Dimension of workpiece | Length, volume angle | Environment machine tool | Static assembly possible |
| Surface geometry | Texture roundness cylindricity etc. | Manufacturing process | Dynamic movement translation and rotation |
| Physical/chemical attributes | Hardness residual stress etc. | Material properties | Determines endurance, wear fatigue |

### *1.4.3   Role of metrology in function* [1.1]

So far, surface metrology has been examined in relation to manufacture. In performance or function it is obviously also important. However, it is necessary to put it into the context of other metrologies.

Table 1.1 shows how the metrologies of dimension, surface and physical properties fit together in function. What it says basically is that the dimensional requirements must be satisfied first, thereby enabling static assembly to be satisfied. Surface metrology enables the dynamic conditions to be tested and is second in importance. If the roundness and straightness as well as the roughness are satisfactory, then it enables the part to rotate in the bearing or to move on a slideway. The final metrology here, called **physical metrology**, includes all non-geometrical aspects such as material properties e.g. hardness, residual stress etc. These, once satisfied, allow long life. The metrologies have to be satisfied in the order shown. Taken as a whole, the table should be named 'Functional metrologies'. The functional aspects of surfaces will be considered in detail in Chapter 5.

## *1.4.4   Standards* **[1.2]**

Table 1.2 gives some of the relevant ISO standards. These standards should always be consulted when specifying surface finish and roundness.

**Table 1.2**   *Relevant ISO standards*

| Terminology | |
|---|---|
| ISO 4287:1997 | Geometrical Product Specifications (GPS) – surface texture: profile method – terms, definitions and surface texture parameters ................................................................ |
| ISO 6813:1985 | Measurement of roundness – terms, definitions and parameters of roundness ....................................................... |
| ISO 8785:1998 | Geometrical Product Specification (GPS) – surface imperfections – terms, definitions and parameters |
| **Properties of surfaces** | |
| ISO 3274:1996 | Geometrical Product Specifications (GPS) – surface texture: profile method – nominal characteristics of contact (stylus) instruments ......................................... |
| ISO 4288:1996 | Geometrical Product Specifications (GPS) – surface texture: profile method – rules and procedures for the assessment of surface texture ...................................... |
| ISO 4291:1985 | Method for the assessment of departure from roundness – measurement of variations in radius |
| ISO 4292:1985 | Methods for the assessment of departure from roundness – measurement by two- and three-point methods ........... |
| ISO 5436:1985 | Calibration specimens – stylus instuments – types, m calibration and use of specimens ....................................... |
| ISO 11562:1996 | Geometrical Product Specifications (GPS) – surface texture: profile method – metrological characteristics of phase correct filters................................................... |
| ISO 12085:1996 | Geometrical Product Specification (GPS) – surface texture: profile method – motif parameters....................... |
| ISO 13565–1:1996 | Geometrical Product Specification (GPS) – surface texture: profile method: surfaces having stratified functional properties – Part 1: filtering and general measurement conditions .............................................. |
| ISO 13565–2:1996 | Geometrical Product Specification (GPS) – surface texture: profile method: surfaces having stratified functional properties – Part 2: height characterization using the linear material ratio curve................................. |

## References

1.1    Reason R. E. *The Measurement of Surface Texture.* Modern Workshop Technology Part 2, ed. Wright Baker, Macmillan, London (1970).

1.2    ISO Handbook *Limits, Fits and Surface Properties.* 2nd edition, ISO, Geneva (1999).

## Bibliography

B1    Bennett J. M. *Surface Finish and its Measurement.* Parts A and B, collected works in optics, Optical Society of America Vol. 12 (1992).

B2    Stout K. J. and Blunt L. eds *Three Dimensional Surface Topography.* 2nd edition, Kogan Page, London (1994).

B3    Thomas T. R. *Rough Surfaces.* 2nd edition, Imperial College Press (1999).

B4    von Weingraber H. and Abou-Aly M. *Handbuch Technische Oberflachen.* Vieweg, Braunschweig (1989).

B5    Whitehouse D. J. *Handbook of Surface Metrology.* Inst. of Phys. Pub., Bristol, Philadelphia (1994).

B6    Whitehouse D. J. *Optical Methods in Surface Metrology.* SPIE Optical Engineering Press Vol. MS. 129 (1996).

# 2

## Identification and separation of surface features

### 2.1  Visualization

The biggest problem with surfaces is that the machining marks are very small. Until the late 1920s, no attempt was made to measure the surfaces. They were examined by eye or by touching with the fingernail. These very subjective tests provided a means of comparison between good and bad surfaces. Providing that these methods were confined to within one factory they could be satisfactory, but if outside checks were needed then such an arbitrary evaluation would be insufficient.

One possibility is the use of standard comparison specimens [2.1]. These are metal plates upon which standard processes such as grinding and turning are represented by a number of roughness values. These standards are made by replication from real surfaces. The surfaces of the specimens are usually chromium-plated for wear resistance. A test surface i.e. a surface under examination can usually be identified by comparing it with the comparison standards by eye and by scratching.

These standards are useful in cases where high accuracy is not needed. Incidentally, some idea of the variety of surface properties can be obtained simply by scratching each specimen with the fingernail. An even better estimate of surface variety can be obtained by viewing the image of a strip light in the standards and varying the angle.

### 2.2  Profiles and roughness – understanding the measurement routine

In order to be able to make sensible estimates of roughness by either stylus or optical methods, it is necessary to define all stages in the measurement procedure.

Perhaps the first requirement is to agree on the nature of the measurement. The simplest starting point is a profile of the surface texture. Unfortunately, this is not straightforward because there are a number of ways of developing a profile.

## 2.2.1  Profiles

This test reveals the potential use of optics in surface measurement.

In order to be able to make sensible estimates of roughness by either stylus or optical methods, it is necessary to define all stages in the measurement procedure. Perhaps the first requirement is to agree on the nature of the measurement. The simplest starting point is a profile of the surface texture. Unfortunately, this is not straightforward because there are a number of ways of developing a profile.

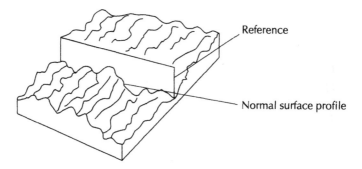

**Figure 2.1 (a)**  *Profile reference planes: plane normal to surface*

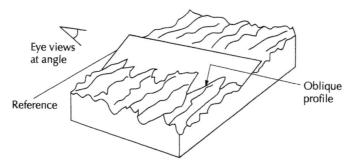

**Figure 2.1 (b)**  *Profile reference planes: plane at angle to surface*

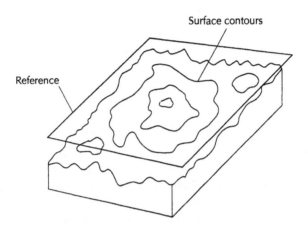

**Figure 2.1 (c)** *Profile reference planes: plane parallel to surface*

In Figure 2.1 (a), the reference is a plane normal to the surface. In Figure 2.1 (b), the reference plane is at an angle and in Figure 2.1 (c) the reference plane is parallel to the general direction of the surface. Oblique profiles are produced if the reference is tilted. Obviously, in Figure 2.1 (b), the tilt of the reference would give a magnified view of surface height, which does not occur in Figure 2.1 (a) [2.2].

Even assuming that the normal profile is preferred, there is still confusion because there are a number of 'normal' profiles. The family of profiles arises from different filtering requirements. These are shown in Figure 2.2.

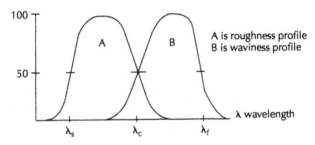

**Figure 2.2** *Profile filters*

$\lambda_s$ profile filter defines the intersection between roughness and even shorter wave components such as microfracture marks.

$\lambda_c$ profile filter defines the intersection between the roughness and waviness profiles at a transmission of 50%. In early standards this was 75%, being a compromise between the British Standard 80% and the US 70.7%.

$\lambda_f$ profile filter defines the intersection between the waviness and even longer wavelength such as form.

Profile filters in general are considered in ISO 115621 [2.3]. These filters produce the following profiles.

1.   *Primary profile ISO 3274*. This is the basis for the evaluation of primary profile parameters and is the profile without filters, apart from stylus effects.
2.   *Roughness profile*. Derived from the primary profile by suppressing the waviness and longer wavelengths by $\lambda_c$.
3.   *Waviness profile*. Derived from the primary profile by suppressing the wavelengths longer than waviness $\lambda_f$ and the shorter wavelengths by $\lambda_c$.

It should be realized that these filters fulfil a purpose. This is to stabilize readings and to focus the measurement on specific detail. Without $\lambda_s$, for example, the measurement would depend on the sharpness of the stylus, which is not necessarily guaranteed by the instrument maker. It could be, for example, that the stylus has become chipped. The filter $\lambda_s$ minimizes the effect of such a possibility.

There are a number of definitions of procedures and parameters that are used extensively. These are contained in ISO 4287 (1997) and ISO 3274.

1.   *Traced profile*. This is the locus of the centre of the stylus tip as the surface is traced. This is the profile from which all other profiles are determined.
2.   *Reference profile*. This is the trace on which the probe is moved within the intersection plane along the guide. It is the practical realization of a theoretically exact profile. Its nominal deviations depend on the deviations of the guide as well as on external and internal system disturbances.
3.   *Total profile*. This is the digital form of the traced profile relative to the reference profile. The total profile is characterized by the vertical and horizontal digital steps (quantization and sample interval).

## 2.2.2   Notation

There are three distances on the surface, which are defined as follows (see Figure 2.3).

1.   *Measurement or sometimes traversing length*. This is the length over which the stylus/pick-up is traversed across the surface. Surface data is not usually taken over all this length. Sometimes, for example, it may be necessary for the traversing mechanism to get up to a constant speed before data logging is started.

2.  *Assessment length*. This is usually shorter than the traversing length and is the length over which surface data is acquired and assessed.

3.  *Sampling length*. This is the fundamental distance over which the surface parameter is assessed, except that in some cases where the available surface is small the sampling length, is smaller than the assessment length. It should not be confused with the sample interval, which is the distance between digital ordinates. It is the smallest distance over which a surface parameter is assessed and is the length over which a reliable estimate of a surface parameter is made. Unless otherwise stated, surface parameters relate to this length. Other uses of the sampling length are given in the account of waviness. See, for example, the International Standard Handbook No. 33 [2.3].

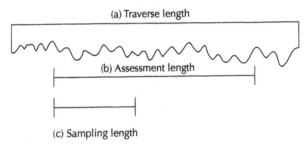

Figure 2.3 *Lengths on surface*

Taking a measurement on a surface would seem to be straightforward. Simply take the surface to the instrument. Run the stylus across the surface and record the meter reading or computer output. This number is then taken to be 'the roughness' of the surface.

Ideally, the operator does not even want to know the 'roughness number'. All that is required is a judgement as to the suitability or otherwise of the surface: anything else is superfluous. Unfortunately, before any such judgement can be made, the validity and fidelity of the reading has to be established. A number of factors have to be decided. Remember that a surface instrument is a calliper in which one arm touches the workpiece surface and the other a reference surface. So, the reference has to be established first before any measurement is taken. In fact, there are two stages in getting the reference. The first step is to ensure that all of the surface signal can be contained within the range of the transducer within the arms of the calliper (Figure 2.4).

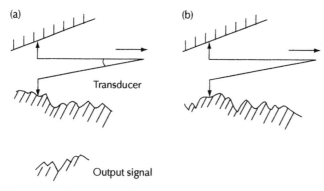

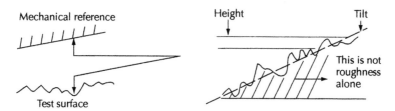

**Figure 2.4** *Calliper system of surface metrology*

Both arms of the calliper have to be kept in contact with their respective surfaces throughout all the measurement. Even then, the signal from the transducer could hardly be called the 'roughness'. The signal output contains irrelevant components of height and tilt (Figure 2.4) concerned with the instrument set-up.

**Figure 2.5** *Assessing roughness: removing tilts and waves*

Obviously, keeping the transducer within range is not all that is necessary. Tilt, height and curves all have to be removed before the roughness can be assessed. There are two ways of doing this. The first is to take the output signal as in Figure 2.5 and compute out tilts etc. by filtering. The other way is to refine the output signal from the transducer by adjustment of the position of the reference surface mechanically. This takes time and so alternative mechanical devices have been employed. These 'skid' methods will be discussed later.

Imagine, therefore, that the output from the calliper has been adjusted either electronically, digitally or mechanically to give a trace suitable to be assessed. Questions arise now as to what length of surface will give a reliable answer.

### 2.2.3  Sampling length – roughness

The key length in Figure 2.6 is the 'sampling length', which has had many names.

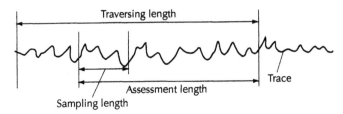

Figure 2.6 *Use of sampling length*

Some of these names are listed below.

1. Cut-off length.
2. Meter cut-off.
3. Filter cut-off.
4. Sampling length – the preferred name.

The 'sampling length' has two roles. One is to include enough surface within it to ensure a reliable value of whichever surface parameter is being used. The other is to preclude waviness – the long wavelength component of the surface signal due to machine tool problems – and just measure the process marks, those produced by machining (the roughness).

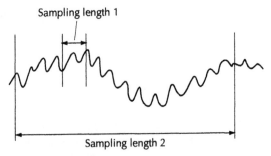

Figure 2.7 *Long and short sampling lengths*

In Figure 2.7, sampling length 1 is too short whereas sampling length 2 is too long. A criterion, which is often applied, extends the sampling length until a visual estimation for a mean line cuts the profile between 20 and 40 times.

It is helpful to delve into the history of the subject to see how this came about. In order to establish a 'typical' sampling length that could be used in industry, a large number of parts were examined at Taylor Hobson. The parts, which numbered tens of thousands, were obtained from Rolls Royce on a shuttle basis in 1940-41. Mr R E Reason of Taylor Hobson found that the most common process was turning, and that the 'average' feed mark was about 0.003 inches. He decided that at least 10 of these feed marks should be included in any sampling length.

Hence, the first sampling length value became 0.003 × 10 = 0.03 inches (or 0.75mm rounded to 0.8mm). Obviously, 10 feed marks within the sample length gives about 20 crossings of the mean line and profile.

To cater for a degree of roughness within the tool feed, the number of crossings was justifiably taken to be up to about 40. This upper limit is also useful for ground surfaces. It is common sense to take only the large detail into account because this usually contributes most to the variability.

Obviously, having only one sampling interval to cover all surfaces is unrealistic. It was decided by Reason and the British Standards in BS 1134 in 1941 to include a sampling length above and below 0.03 inches (0.8mm). This could cater for milled surfaces (upper cut-off) and fine grinding (shorter cut-off). Originally, the ratio between sampling length values was to give a decade change i.e. 10:1 in two sampling length changes. This number was soon rounded off to the more convenient but arbitrary 3.2:1. Nowadays, the sampling length value can be less rigidly fixed. The essential point is to qualify each measurement of the surface with the cut-off value, otherwise comparison between surface readings is likely to be wrong.

Also, it is usual to arrange five sampling lengths within one assessment length. At one time the ratios were different, namely three for the long cut-off 0.1" (2.5mm), five for the 0.03" (0.8mm) and seven for 0.01" (0.25mm). As the ratio determines the variability of the parameter, it is not surprising that the shorter cut-offs were more consistent. To avoid confusion, most of the surface finish parameters refer to one sampling length in the definition so that the answer for all instruments should be the same. In practice, however, the default number of sampling lengths is five, but this number should accompany the parameter value on the drawing.

The concept of the sampling length is very useful for conventional processes but there is sometimes a problem if the surface has been generated by more than one process, all of which will have left remnants in the final profile.

In this circumstance, the process that dominates the profile should be used in the criterion if possible. The problem is that the different components of the profile want to be treated individually but cannot be because they are all trapped within the same profile. Such multiprocess surfaces pose problems in evaluation. Figure 2.8 shows the profile of one such multiprocess – plateau honing.

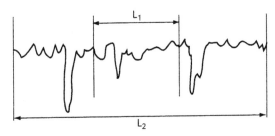

**Figure 2.8** *A multiprocess, plateau honing process*

In this example, picking the sampling length $L_1$ based on the surface roughness of the 'plateau' would be wrong because parameters would be dramatically altered if just one deep valley got included. Under these circumstances, a much longer sampling length should be used such as $L_2$ in which a number of deep valleys would always be included within the sampling length. This conclusion is not reached because of variability alone. Another reason is that the presence of just one scratch can distort the remainder of the profile. Extraordinary techniques have to be used to remove this effect. It is simpler to increase the sampling length. In typical plateau honing, this would entail moving from 0.8mm (0.03″) to 2.5mm (0.1″).

## 2.3   Waviness

The removal of waviness from the profile – by means of the sampling length – has always been contentious. Some argue that it is only another wavelength and should be included in the roughness assessment; others say that it should not be included. There are two aspects to this. Is the waviness important for monitoring the manufacture and is the waviness important – in its own right – functionally?

If waviness is defined as a characteristic produced by an imperfect machine tool then it should be distinguished from roughness. There is no doubt of its relevance in this case.

Functionally, the situation is more complicated because the relevance of the waviness depends on the nature of the workpiece.

Contact behaviour is crucial to most industrial applications. Here, waviness is important. Furthermore, it is different from the roughness. The key to this is the method of generation. Figure 2.9 shows that waviness is attached to low-energy generation whilst roughness is determined by high-energy processes. The effects of roughness generation on material properties underneath the surface are evident at very small depths $\sim 0.5\mu m$. This is because the maximum surface stress position is determined by the local roughness geometry e.g. the curvature. The same is true for waviness only at much greater depths and much smaller strain because of the longer wavelengths. The resultant effect is shown in Figure 2.10. Waviness and roughness have different effects and so should be separated in contact situations (Figure 2.11). It is not necessary to separate them in non-contact functions such as in light scatter. If in doubt, measure them separately. They can always be reconstituted.

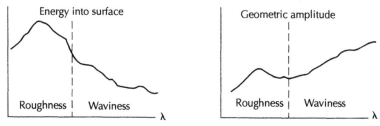

**Figure 2.9** *Energy into surface vs. amplitude*

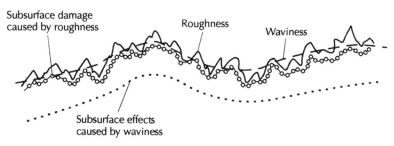

**Figure 2.10** *Stress profile vs. geometric profile*

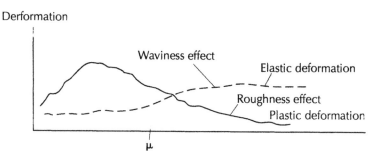

**Figure 2.11** *Mode of deformation*

From what has been said, it might be thought that the concept of 'sampling length' is confined to roughness measurement in the presence of waviness. Historically, this is so. Recent thinking has suggested that in order to rationalize the measurement procedure, the same 'sampling length' procedure can be adopted to measure 'waviness' in the presence of form error, and so on to include the whole of the primary profile. Hence $l_r$, the sampling length for roughness, is joined by $l_w$. For simplicity, roughness is usually the surface feature considered in the text.

## 2.4   Implementing the concept of sampling length

### 2.4.1   General

This might seem to be a trivial task with today's computers. Historically, it has been a nightmare. There are a number of requirements that conflict. Originally, about 1939 chart recorders were incorporated into instruments – in particular the Taylor Hobson Talysurf range. The idea was to use the chart to check the meter. The problem was to decide the primary method – the meter or the chart. Was the meter reading an approximation to the chart or vice versa? Another complication arose in deciding the basis for the graphical method. The UK and USA opted for a mean line 'M' method based on a best fit line, whereas Professor Von Weingraber in Hannover, Germany, advocated an envelope called the 'E' system (Figure 2.14) based on a rolling ball. The E system is not trivial. It is based on sound engineering, as will be seen.

### 2.4.2   The values

The sampling length has to be chosen for different processes as well as for different filters. The choices are to some extent arbitrary but they are given below to provide a useful framework. Table 2.1 gives a Gaussian filter selection. Other filters that are non-standard have different weighting functions. Tables 2.2, 2.3 and 2.4 give the sampling lengths for different parameters.

**Table 2.1**   *Gaussian filter selection [1.2]*

| Gaussian Filter Sampling Lengths for $R_a$, $R_z$ & $R_y$ of periodic profiles | | | 1995 |
|---|---|---|---|
| $S_m$ mm | | | |
| Over | Up to (inclusive) | Sampling length mm | Evaluation length mm |
| (0.013) | 0.04 | 0.08 | 0.4 |
| 0.04 | 0.13 | 0.25 | 1.25 |
| 0.13 | 0.4 | 0.8 | 4.0 |
| 0.4 | 1.3 | 2.5 | 12.5 |
| 1.3 | 4.0 | 8.0 | 40.0 |

In Tables 2.2, 2.3 and 2.4 a notation will be used, which is explained fully later. See glossary at the end of this book.

**Table 2.2**  *Roughness sampling lengths for the measurement of $R_a$, $R_q$, $R_{sk}$, $R_{ku}$, $R_{\Delta q}$ and curves and related parameters for non-periodic profiles (for example ground profiles)*

| $R_a$ | Roughness sampling length *lr* | Roughness evaluation length *ln* |
|---|---|---|
| μm | mm | mm |
| $(0.006) < R_a \leqslant 0.02$ | 0.08 | 0.4 |
| $0.02 < R_a \leqslant 0.1$ | 0.25 | 1.25 |
| $0.1 < R_a \leqslant 2$ | 0.8 | 4 |
| $2 < R_a \leqslant 10$ | 2.5 | 12.5 |
| $10 < R_a \leqslant 80$ | 8 | 40 |

**Table 2.3**  *Roughness sampling lengths for the measurement of $R_z$, $R_v$, $R_p$, $R_c$ and $R_t$ of non-periodic profiles (for example ground profiles)*

| $R_z$ [1]  $R_{z1}$max. [2] | Roughness sampling length *lr* | Roughness evaluation length *ln* |
|---|---|---|
| μm | mm | mm |
| $(0.025) < R_z, R_{z1}$max. $\leqslant 0.1$ | 0.08 | 0.4 |
| $0.1 < R_z, R_{z1}$max. $\leqslant 0.5$ | 0.25 | 1.25 |
| $0.5 < R_z, R_{z1}$max. $\leqslant 10$ | 0.8 | 4 |
| $10 < R_z, R_{z1}$max. $\leqslant 50$ | 2.5 | 12.5 |
| $50 < R_z, R_{z1}$max. $\leqslant 200$ | 8 | 40 |

1) $R_z$ is used when measuring $R_z$, $R_v$, $R_e$ and $R_t$
2) $R_{z1}$max. is used only when measuring $R_{z1}$max., $R_{v1}$max., $R_{p1}$max., and $R_{e1}$max.

**Table 2.4**  *Roughness sampling length for the measurement of R-parameters of periodic profiles, and $RS_m$ of periodic and non-periodic profiles*

| $RS_m$ | Roughness sampling length *lr* | Roughness evaluation length *ln* |
|---|---|---|
| mm | mm | mm |
| $0.013 < RS_m \leqslant 0.04$ | 0.08 | 0.4 |
| $0.04 < RS_m \leqslant 0.13$ | 0.25 | 1.25 |
| $0.13 < RS_m \leqslant 0.4$ | 0.8 | 4 |
| $0.4 < RS_m \leqslant 1.3$ | 2.5 | 12.5 |
| $1.3 < RS_m \leqslant 4$ | 8 | 40 |

### 2.4.3   Basic rules for the determination of cut-off wavelength for the measurement of roughness profile parameters [1.2]

When the sampling length is specified in the requirement on the drawing or in the technical product documentation, the cut-off wavelength, $\lambda_c$, is then the sampling length. When no roughness specification exists or the sampling length is not specified in a given roughness specification, the cut-off wavelength is chosen by procedures given below.

## Measurement of roughness profile parameters

When the direction of measurement is not specified, the workpiece should be positioned so that the direction of the section corresponds to the maximum values of height of the roughness parameters $(R_a, R_z)$. This direction is normal to the lay of the surface being measured. For isotropic surfaces, the direction is arbitrary.

Measurements should be carried out on that part of the surface that is judged to be critical. This can be assessed by visual examination. Separate measurements should be distributed equally over this critical part of the surface to obtain independent measurement results.

To determine roughness profile parameter values, first view the surface and decide whether the roughness profile is periodic or non-periodic. Based on this determination, one of the procedures specified below have to be followed unless otherwise indicated. If special measurement procedures are used, they must be described in the specifications and in the measurement protocol.

## Procedure for non-periodic roughness profile

For surfaces with a non-periodic roughness profile, the following procedure is followed. It is to some extent an iterative procedure.

(a)  Estimate the unknown roughness profile parameter $R_a$, $R_z$, $R_{z1}$ max. or $RS_m$ by any means preferred, for example visual inspection, roughness comparison specimens, graphical analysis of a total profile trace etc.

(b)  Estimate the sampling length from Tables 2.1, 2.2 or 2.3, using the $R_a$, $R_z$, $R_{z1}$ max. or $RS_m$ estimated in step (a).

(c)  With a measuring instrument, obtain a representative measurement of $R_a$, $R_z$, $R_{z1}$ max. or $RS_m$, using the sampling length estimated in step (b).

(d)  Compare the measured values of $R_a$, $R_z$, $R_{z1}$ max. or $RS_m$ in Tables 2.2, 2.3 or 2.4 corresponding to the estimated sampling length. If the measured value is outside the range of values for the estimated sampling length, then adjust the instrument to the respective higher or lower sampling length setting indicated by the measured value. Then measure a representative value using this adjusted sampling length and compare again with the values in Tables 2.2, 2.3 or 2.4. At this point, the combination suggested by Tables 2.2, 2.3 or 2.4 of the measured value and the sampling length should have been reached.

(e)  Obtain a representative value of $R_a$, $R_z$, $R_{z1}$ max. or $RS_m$, for one shorter sampling length setting (if this shorter sampling length setting has not previously been evaluated in step (d). Check to see if the resulting combination of $R_a$, $R_z$, $R_{z1}$ max. or $RS_m$ and sampling length is also that specified by Tables 2.2, 2.3 or 2.4.

(f)  If only the final setting of step (d) corresponds to Tables 2.2, 2.3 or 2.4 then both the sampling length setting and the $R_a$, $R_z$, $R_{z1}$ max. or $RS_m$ value are correct.

(g)  Obtain a representative measurement of the desired parameter(s) using the cut-off wavelength value (sampling length) estimated in the preceding steps.

## Procedure for periodic roughness profile

For surfaces with a periodic roughness profile, the following procedure is used.

(a)  Estimate graphically the parameter $RSm$ of the surface of unknown roughness.

(b)  Determine the recommended cut-off wavelength value for the estimated parameter $RS_m$ using Table 2.3.

(c)  If necessary e.g. in cases of disputes, measure the $RS_m$ value using the cut-off wavelength value determined according to (b).

(d)  If the $RS_m$ value from step (c) relates, according to Table 2.3, to a smaller or greater cut-off wavelength value than in step (b), use the smaller or greater cut-off wavelength value.

(e)  Obtain a representative measurement of the desired parameter(s) using the cut-off wavelength value (sampling length) estimated in the preceding steps.

So far, the extent of the sampling length has been discussed. It has been proposed in accordance with ISO that $S_m$ is a suitable parameter upon which to base the estimation. This is obvious because $S_m$ – the average distance between major peaks – is an estimate also of the size of the 'independent' events contained within the profile and which ultimately determine the reliability of the parameter reading. In fact, $S_m$ is just over three times the independence length for a random surface.

One other issue is the shape and basis of the reference line used within the sampling length. As previously stated, the 'M' system was advocated by the UK, the 'E' or envelope system by Germany and there is another based on the 'motif' system, which has been advocated for many years by the French. None of these are right and none are wrong. The motif method is based upon envelopes of the peaks and valleys. This will be considered later. So, the reference shapes are based on the M system – all profile, the E system – peaks, or the motif system – peaks and valleys.

## 2.5   The shape of the reference line

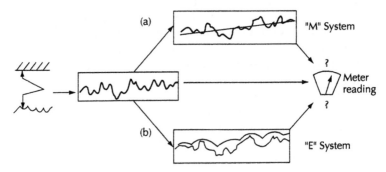

**Figure 2.12**   *Use of 'E' or 'M' system*

### 2.5.1   Envelope system

As will be seen later in Section 2.7, the British approach (Figure 2.12) was pragmatic and was an attempt to get chart and meter agreement. The German approach was more functional. In it, a ball was rolled across the profile. The locus of the centre of the ball was dropped to touch the profile. This line was then taken as reference (Figure 2.13).

The choice of a ball or circle (for the profile) to determine the reference was based on two considerations. One was functional in the sense that the ball was meant to simulate the largest reasonable radius at a contact i.e. peak curvature. This fixed the value more or less by experience. That there was a considerable argument over a 25 or 2.5mm radius is evidence of the woolly definition. There was also, however, another more practical estimate based on using a micrometer to measure the diameter of the workpiece.

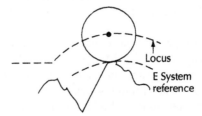

**Figure 2.13**   *E reference line system*

The idea was to link surface metrology to dimensional metrology. In other words, the reference for the surface metrology was in fact the basis for the dimensional measurement of size by using a micrometer (Figure 2.14).

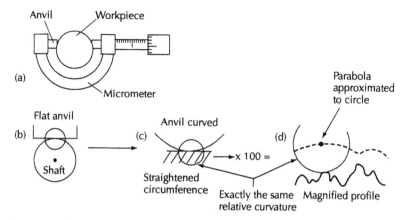

**Figure 2.14**  *Use of micrometer for measurement of dimensions*

The circle or ball diameters were 25mm and 2.5mm.

One basic problem with the E system is that it is not linear. There are two mechanisms: one where the circle spans the crests and the other when it bottoms. (See Figure 2.16).

The elegant German method fell down because although it was possible to simulate on a chart, it was difficult if not impossible in the early days to instrument. Attempts to generate the envelope at the mechanical input, i.e. literally to have a ball stylus with the sharp stylus poking through it, failed for mechanical reasons i.e. dirt (Figure 2.15).

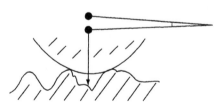

**Figure 2.15**  Circle spanning crests

(a)

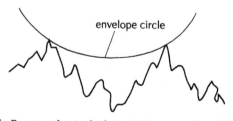

**Figure 2.16**  *The E systems showing bridging and bottoming (a) envelope bridges peaks*

(b)

(c)

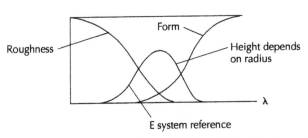

**Figure 2.16 (continued)**   *The E system showing bridging and bottoming (b) envelope bottoms and (c) E system transmission curve*

The E system failed because the meters in use were better suited to measuring average parameter values of the roughness rather than peaks to which the E system was suited. Getting rid of conventional meters and replacing them with computers was at the time impractical.

Unfortunately, linking dimensional and surface metrology is difficult even now because the requirements for range and resolution are different. It needs instrumentation with a range suitable for dimensional metrology and a resolution suitable for surface metrology – a point considered later on.

Trying to get an optimum system is plagued with what appears to be irrelevant constraints. For example, in order to safeguard the meter from measuring set-up problems such as tilt, dc and curvature signals, an electrical filter was inserted before the meter. This was a two stage 2CR filter.

In the first instance, the insertion was intended to minimize skid effects but eventually turned out to be useful as a filter to implement the sampling length. Unfortunately, this filter insertion made matching of the graphical reference with the meter very difficult. Although this 2CR filter was not ideal, it has been incorporated into all the early surface roughness standards e.g. BS 1134 and B46. Nowadays, it can be circumvented but many old instruments still use it so it cannot be ignored.

Before this use of filters is discussed, some graphical ways will be mentioned. These are not irrelevant. It is conceivable that users still do not have computer-based instruments and yet still have to calibrate the output. Such users have to be catered for although it clutters up the standards documentation.

Should anyone think that the arguments about reference lines are of marginal importance, it should be remembered that fitting a reference to the data (i.e. placing a suitable line within the collected data) is *the very basis* of surface metrology. This reference does not exist so it has to be constructed. Getting this wrong invalidates the whole process of surface metrology. This is why there have been so many attempts at producing a suitable procedure.

So, the position in space and the shape of the reference are crucial. Adjusting the level and tilt of the instrument to get the specimen within range of the transducer only provides the first step – mechanical reference. There is absolutely no guarantee that this mechanical reference lies within the profile. When setting up, it is usually remote. It is the subsequent processing that positions the reference within the profile.

Particularly vulnerable is the estimation of waviness. The waviness line – the mean line, is contrived. It has to be constructed and evaluated from either the roughness or form geometry.

### 2.5.2   Motif methods

The motif method is an envelope technique. It was suggested in France in the early 1970s; see Standard P.R.E.05-015 (1981). Its principal exponent was Professor Biele who spent a great deal of time implementing the system into the Peugeot automobile plant and M. Schaeffer who converted it into industrial practice. The system in those days was called '$R$ and $W$'. In this method, a line is drawn along the peaks according to a given rule and a line through the valleys. The roughness value $R$ is then the average of 10 measurements of roughness between the 'envelope superior' (the peak envelope) and the 'envelope inferior'. The waviness was obtained by applying envelope rules to the peak envelope.

The 'motif' method in principle is rather simple (see Figures 2.17, 2.18 and 2.19). The unfiltered profile is divided into useful geometrical characteristics, **motifs**, in accordance with an empirically found algorithm that should be determined by the combined experience of people dealing with profile analysis. In one form, the motif itself is two profile peaks with a valley in between them.

### Motif reference

This technique is different from that using the sampling length. It is more of an 'intrinsic' reference in which the reference line is found from the behaviour of surface peaks. Because of this, 'peaks' have to be defined. These will be given below for motif generation but are also valid for general surface metrology.

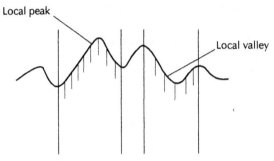

**Figure 2.17**   *Local peak and valley*

Local peak     – part of the profile between two adjacent valleys,
Local valley   – part of the profile between two adjacent peaks,
Motif          – a portion of the primary profile between the highest points of two
                 local peaks of the profile that are not necessarily adjacent.

A motif is characterized by the following features.

(a)  $AR_i$ or $AW_i$, the **length** measured parallel to the general direction of the profile. $R$ is the roughness and, alternatively, $W$ the waviness.
(b)  Two depths $H_j$ and $H_{j+1}$ or $HW_j$, $HW_{j+1}$ measured perpendicular to the general direction of the primary profile.
(c)  A characteristic called $T$, which is the smallest depth between the two depths shown.

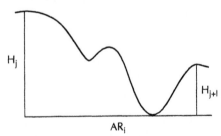

**Figure 2.18**   *Motif of roughness*

In Figure 2.19, $T = H_{j+1} = \mathrm{MIN}[H_j, H_{j+1}]$.

From these, the **roughness motif** is derived using the above characteristics but subject to a limiting value A, where A is some value determined to include very small deviations.

**Figure 2.19**   *Upper envelope line*

## Waviness motif

This is derived from the upper envelope line using the operator above with a limit value B.

From Figure 2.20, $T = H_{wj+l}$.

## Parameters

Mean spacing of roughness motifs $AR$.

$$AR = \frac{1}{n \sum_{i=1}^{n} AR_i}$$ (n is the number of $AR_i$ values with mean depth of roughness

motifs $R$),

$$R = \frac{1}{m \sum_{j=1}^{m} H_j}$$ (m is the number of $H_j$ values).

where the number of $H_j$ values is twice that of the $AR_i$ values (m = 2n).

## Maximum depth of profile irregularity $R_x$

This is the largest depth $H_j$ within the evaluation length.

**Note:** waviness evaluation is exactly similar where the upper envelope line is the effective profile.

For ways in which motifs are combined, see ISO 12085 1996 (E) information.

The motif is characterized by the two heights $H_1$ and $H_2$. The characteristic depth of the motif has a special significance in the evaluation: it is the smaller of $H_1$ and $H_2$; the mean depth is $(H_1 + H_2)/2$ and $AR_1$ is the distance between the peaks.

Application of the motif method involves checking whether two adjacent motifs can be treated separately or whether one of the motifs can be regarded as insignificant and absorbed into the other one to make one large motif. Two motifs cannot be merged if any of the four conditions below apply.

1.  Envelope condition (Figure 2.20 (b)(ii)). Two adjacent motifs cannot be combined if their common middle peak is higher than the two outer peaks.
2.  Width condition (Figure 2.20 (b) (iii)). Adjacent motifs can only be combined if the result is less than or equal to 500µm. This is a very weak condition, indicating that motifs of up to 500µm are to be included as roughness. It corresponds in effect to a cut-off. The width limit for evaluating waviness is 2500µm. These arbitrary rules have been proposed with the car industry in mind.
3.  The magnification condition (Figure 2.20 (b) (iv)). Adjacent motifs may not be joined if the characteristic depth of each of the adjacent depths $T_j$ of the result m is less than the largest characteristic depth of each of the adjacent motifs. This condition is most important. It means that in each case an unequivocal evaluation can be reached independently of the tracing direction.
4.  The relationship condition (Figure 2.20 (b)(v)). Adjacent motifs may be considered if at least one has a characteristic depth less than 60% of the local reference depth $T_R$. Thus, adjacent motifs of approximately the same size cannot be combined to form one single overall motif.

## Motif procedure (according to Fahl [2.6], Figure 2.20(b))

Find the local reference depth over the relevant sections in the profile. Then, within each individual section, every single motif is checked against its neighbour using the four conditions above and whenever necessary combined with a new motif.

Every new motif must be compared with its neighbour until no further combination of two motifs is possible within that section. This is carried out for each section of the profile. When completed, all motifs that lie outside the limits of the section must be checked with each other until no further mergers are possible.

In theory, this is the end result. In practice, however, there are other considerations. Examples of this include the presence of single isolated peaks or valleys. These have to be smoothed to a 'reasonable' level. After this, the 'corrected' profile results. Then $R_m$ the average motif depth, $A_R$ the average width, and $p(R_m)$, $p(A_R)$ the distribution of $R_m$ and $A_R$, are found. After these calculations, a near profile comprising the peaks just found is made. The whole procedure is repeated except that, instead of 500µm, 2500µm is used. This corresponds to the envelope above the corrected profile. This time, $W_m$ and $A_w$ and $p(W_m)$, $p(A_w)$ are found and taken as the waviness parameters. The whole procedure can be repeated to give the form profile.

This represents the complete breakdown of the profile into roughness, waviness and form.

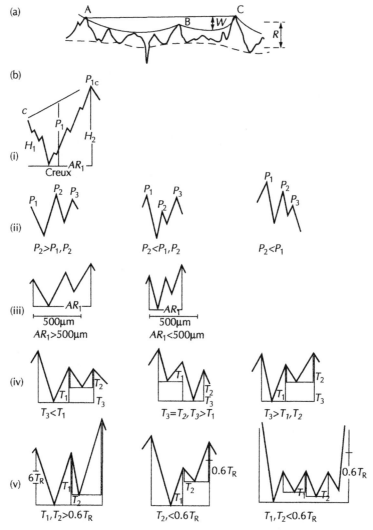

**Figure 2.20** *(a) Motif method R and W – (b) (i) Definition of motif (ii) Envelope condition (iii) Width condition (iv) Magnification condition (v) Relationship condition*

The reference depth $T_R$ may appear in two forms. It may be the largest depth found in any one suitable profile section approximately 500µm wide; it is then called the **local depth reference**. Alternatively, it may be the characteristic depth $T_j$, which a combined motif would have.

In this motif method, $R_m$ is approximately $R_y$ or $R_z$ qualified by a varied sampling length. $A_R$ or AR is the average width between significant peaks. This corresponds to $S_m$, the average distance between zero crossings of the profile with the mean line.

The advantage is that the rules are arbitrary. They could, in principle, be modified to suit any given function. Another advantage is that the motifs are peak oriented; quite major peaks near to a valley are ignored, yet small peaks near to a major peak are not. A different set of rules would be needed to make the algorithm symmetrical. The arbitrary selection of 500mm and 2500mm is a disadvantage but no more than the 0.8mm cut-off selected for the filters.

It seems that the motif method is another way of looking at profiles, perhaps complementing the filtering methods or perhaps best suited for dealing with multiprocess surfaces where no standard characteristics occur.

Another possible use is in plotting the distribution of given sizes of motifs. This is almost the metrology equivalent of a Pareto curve. Using this curve motif size against frequency of occurrence can be plotted for roughness and waviness and can represent some interesting information. Exactly how it can be used is not obvious but it is a different type of breakdown, which does not rely on a reference line but on adequate definitions of peaks.

Motif methods are unusual in the sense that specific parts of the profile determine the reference rather than all of it as in filtering. This aspect has its advantages and its disadvantages. It is focused – the motif operation is flexible but the very fact of having peaks and valleys as the reference and to a large extent the functional parameter make them very susceptible to high frequency effects.

## 2.6   Other methods

### 2.6.1   Best-fit straight line

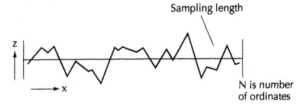

**Figure 2.21**   *Best-fit line*

This line is positioned such that the sum of the squares of the ordinate values above and below it are a minimum (Figure 2.21).

$$\text{Slope} = m = \frac{\sum x \sum z - N \sum xz}{(\sum x)^2 - N \sum x^2}$$

where x is the horizontal distance along the surface and the z value the vertical distance. This method is unique but it can give some strange results (see Figure 2.22).

## Intrinsic filtering

The representation of the original profile in blocks is reminiscent of the methods of filtering using the sampling length described earlier. In this case, however, the blocks are determined from the surface itself and could therefore better be described as 'intrinsic filtering'. Any local effects on the surface that are not characteristic will automatically be taken into account. The sampling 'matches' itself to the local properties of the surface.

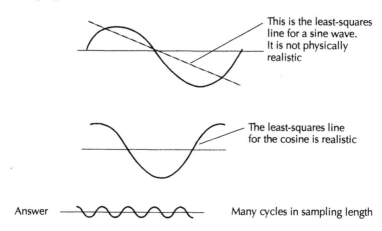

This is the least-squares line for a sine wave. It is not physically realistic

The least-squares line for the cosine is realistic

Answer

Many cycles in sampling length

**Figure 2.22**   *Problems arising in creating a best-fit line*

## 2.6.2  Polynomial fitting

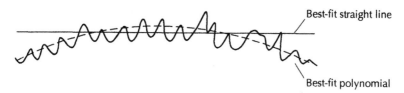

Best-fit straight line

Best-fit polynomial

**Figure 2.23**   *Best-fit curve*

The best-fit polynomial is used effectively if the order of the curve is known (Figures 2.23 and 2.24). It can cause problems if a given curve is forced upon the unknown one.

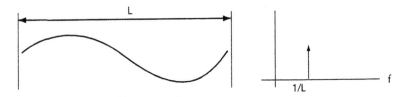

**Figure 2.24**   *Fitting Fourier coefficients to a sine wave*

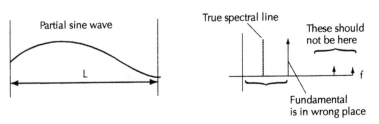

**Figure 2.25**   *Fitting Fourier coefficients to a partial wave*

Consider the case shown in Figures 2.24 and 2.25 of fitting Fourier coefficients to a sine wave. If the wave is complete, the single line spectrum is correctly determined but if only a partial wave as in Figure 2.25 is available, the spectrum is highly suspect.

## 2.7   Filtering and M system

Deciding on the sampling length value is one aspect of filtering. The other is the method of achieving it. Here, methods involving filtering will be discussed. These are not usually graphical because the amount of calculation on the chart is extremely high.

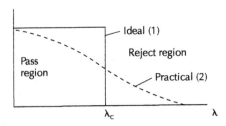

**Figure 2.26**   *Determination of mean line within sampling length*

The sampling length determines $\lambda_c$ but there are many ways of having some acceptance in region 1 and some rejection in region 2. The shape of the accept/reject curve determines the characteristic of the mean line within the sampling length (see Figure 2.26).

### 2.7.1   Mean lines – terminology

The assessment of the surface texture involves blocking certain wavelengths and accepting others.

Consider the roughness, for example.

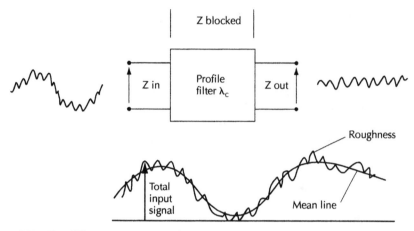

**Figure 2.27**   *Use of filtering in assessing surface texture*

The blocked signal in the case of roughness is the waviness and form (Figure 2.27). This constitutes a mean line from which the roughness is assessed. In the UK system, the 'mean line' used to be called the 'centre line'. They are the same!

A major problem in surface metrology has been the procedure determining the shape and length of the mean line. By definition, the area subtended between the profile and the mean line has to have an equal value above and below, but does not necessarily have to satisfy a 'best-fit' least-squares criterion. For example, a mean line found by using an envelope or motif reference line as in Section 2.5 is not a best fit in terms of least squares.

### Mean line terminology
1.   Mean line for roughness profile – line corresponding to the long wavelength component suppressed by the profile filter $\lambda_c$ (See ISO 11562 1996,3,2).
2.   Mean line for waviness profile – line corresponding to long wavelength component suppressed by the profile filter $\lambda_f$ (ISO 11562. 1996).
3.   Mean line for primary profile – line determined by fitting a least-squares line of nominal form through the primary profile.

### Sampling length terminology

It is obvious from the above that the term 'sampling length' can be applied to waviness and form in exactly the same way as they were considered for roughness discussed in Section 2.1. Hence the general definition and notation for sampling length.

General sampling length $l_p$, $l_r$, $l_w$ – length in the direction of the x axis used for identifying the irregularities characterizing the profile under evaluation (ISO 4287, 1997(E/F)).

**Note**: the sampling length for the roughness $l_r$ and the waviness $l_w$ is numerically equal to the characteristic wavelength of the profile filters $\lambda_c$ and $\lambda_f$ respectively. The sampling length for the primary profile $l_p$ is equal to the evaluation length. In other words, in the case of the primary profile there can only be one sampling length whereas for the roughness and waviness there can be a number (usually five in the case of roughness) within the evaluation length.

## 2.7.2  Filtering – general

The term 'filtering' covers all ways of separating different wavelengths. A filter can be conventionally made up from electrical components like capacitors and resistors, or it can be a digital filter working with an algorithm giving the same effect as the electronic filter. Alternatively, polynomials can be used to determine the mean line. The advantages and disadvantages of using a filter are:

1.   It is a linear operator on the profile data irrespective of the nature of the data. No prior knowledge of the data is needed.
2.   It is easy to change the characteristics of the filter by moving into the frequency domain.
3.   It loses data at the beginning and end of the assessment length (Figure 2.28).

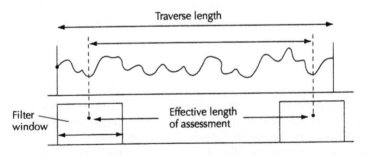

**Figure 2.28**   *End effect of using a filter*

The process of filtering is shown below. There are two ways of operating, one in space and the other in frequency (or wavelength). Note that surface metrology utilizes filtering in space, not time.

*Path A* is all in the space domain i.e. all in *x*, the distance along the profile. This path requires a convolution at step 1, which is a window scanning across the profile. This requires a lot of multiplications and additions just to get one filtered point but requires only one basic step.

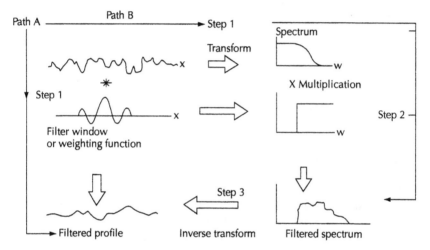

**Figure 2.29** *Filtering using convolutional transformation*

*Path B* is in the frequency or wavelength domain. This requires three steps as opposed to one. Two of the steps are transformations and one is a multiplication. The point here is that transformations by the FFT routine is quick compared with a convolution operation.

Both paths are equivalent. In surface metrology, especially in roughness, it is more visual to use Path A because a workpiece exists in space, not frequency. Path B can be used effectively in roundness because the variable is periodic.

**Note**: convolution is a mathematical operation that is similar to looking at the profile through a window (Figure 2.30). The output at any position is the total light coming through the window. The width of the window is equivalent to the sampling length and the shape of the window depends on the type of filter. The symbol for convolution is *. Figure 2.29 shows that there are always two ways of achieving the filtering: one direct – using convolution, and the other using transformations.

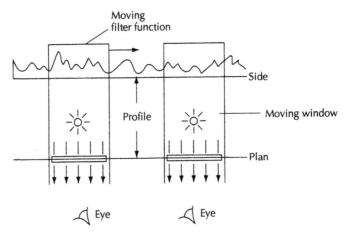

**Figure 2.30** *Convolution as a window*

### 2.7.3 Standard electrical 2CR filter

The first filter to be used was the 2CR filter shown in Figure 2.31. This filter was originally used to get rid of tilt and curvature on skid instruments but soon became the accepted waviness filter. The long wavelengths are blocked by the capacitors, which only allow small waves through. At the time, this filter seemed adequate but it soon became clear that if peak-like parameters were to be measured, this filter was unacceptable. Research at Taylor Hobson revealed the problem and produced the solution, which has now belatedly been broadly adopted [2.9].

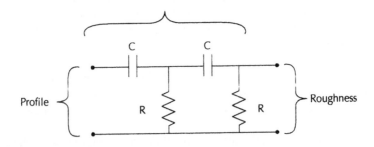

**Figure 2.31** *Behaviour of an analogue filter*

Because the 2CR filter is analogue and works in time rather than space, it has an asymmetrical window (or weighting function), which produces a distorted waviness mean line shown in Figure 2.32. Also, the attenuation characteristics proved to be too gradual as illustrated in Figure 2.32. A sharper cut was called for. Incidentally, it is not sensible to use a filter that has an infinitely sharp cut-off as shown in Figure 2.26. Common sense dictates that a change in wavelength of a surface by a small amount cannot affect the performance dramatically.

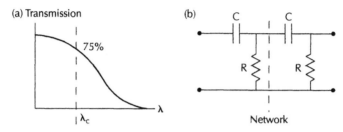

**Figure 2.32**    *Characteristic of a standard filter (2CR)*

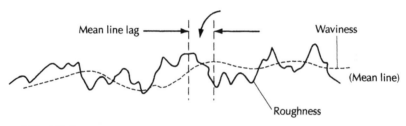

**Figure 2.33**    *2CR distortion*

### 2.7.4 Phase corrected filters

It has been explained why the analogue 2CR filter was used in early instruments. This filter helped considerably in instrument set-up and was widely adopted. Unfortunately, it introduces some undesirable effects, as seen above.

The solution to this distortion was given by Whitehouse at Taylor Hobson as early as 1965. It can be explained by referring to Figure 2.34. The window function in its simplest form is literally a window. The output is simply the average value of the profile enclosed by the window. Other shapes are possible but in order to be distortion free, they must be symmetrical about an x value. Asymmetry produces mean line distortion. Thus, Figure 2.34 (a) shows the mean line weighting (or window function) for the conventional 2CR filter. Figure 2.34 (b) shows the weighting function for the same transmission characteristics, yet having no distortion because it is symmetrical about $x_0$ [2.10].

(a) Asymmetrical          (b) Symmetrical

**Figure 2.34**  *Effect of symmetry on distortion*

There are many shapes for the weighting function (sometimes called **window function**) but the one now accepted as standard is the Gaussian shape shown in Figure 2.35 [2.11].

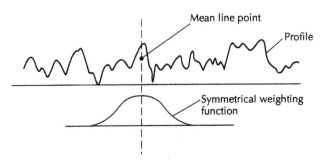

**Figure 2.35**  *Weighting function*

This is a rather well-behaved bell shape. It is well to remember that this shape is just convenient. It has no metrological merit other than the fact that the shape is well known in quality e.g. in statistical process control. It can also be generated in a computer.

These distortionless filters have adopted the name originally suggested by Whitehouse. They are called **phase corrected filters** (also known as **linear phase filters**).

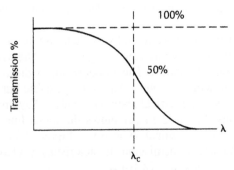

**Figure 2.36**  *Phase corrected transmission filter*

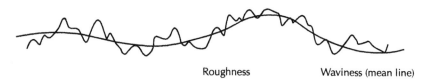

Roughness                    Waviness (mean line)

**Figure 2.37**   *Phase corrected filter*

The attenuation characteristic has also been fixed by agreement with ISO members. For a detailed explanation of these filters, consult [2.10].

## 2.8 Conclusions

This chapter has demonstrated the complex and not altogether logical evolution of methods to isolate roughness from the profile. It is not surprising that the development has been somewhat uneven. It should be remembered that the people needing the roughness for control purposes were production engineers not used to waveform analysis as were electrical engineers. So, terms such as cut-off and bandwidth were to some extent unfamiliar. Having a cut-off of 75% for the 2CR filter was expedient; bridging the values of the UK and USA, but hardly scientific. There are other instances in surface metrology where similar arbitrary values turn up. These are inevitably embedded in the standards system quickly to push forward quality of conformance. With hindsight, they could well have been left out. Unfortunately for the instrument maker, the old as well as the new have to be catered for. This has resulted in a bewildering variety of filters and parameters.

## References

2.1   Hillman W., Klanz O. and Eckol T. Wear Vol. 97 p27 (1984).

2.2   Schmalz G. Z. VDI Vol. 73 p144 (1929).

2.3   ISO Handbook 33 *Limits, Fits and Surface Props.* Geneva (1998).

2.4   Reason R. E. et al *Report on the Measurement of Surface Texture by Stylus Methods.* Rank Organization (1944).

2.5   von Weingraber H. *Uber die Eignung der Hullprofils als Bezugslinie fur Messung der Rauheit. Microtechnic* Vol. 11 p6–11 (1957).

2.6   Fahl C. *Handschriflicher.* Ber. Verfasser Vol. 28 p6 (1982).

2.7   Standard R. R. E. 05-1015 (1981).

2.8   ISO 12085, (1996E).

2.9   Whitehouse D. J. and Reason R. E. *Equation of Mean Line of Surface Texture Found by Electric Wavefilter.* Rank Org (1965).

2.10  Whitehouse D. J. *An Improved Wavefilter for Use in Surface Texture.* Proc. Inst. Mech. Eng. Vol. 182 pt 3k p306 (1967).

2.11  Bodschwinna H. and Seewig J. XI Inty. Coll. On Surfaces Chemnitz, Jan (1966).

# 3

## Profile and areal (3D) parameter characterization

### 3.1 Specification

Before embarking upon the parameter evaluation, some terms have to be established:

> $R$   is roughness profile,
> $W$  is waviness profile,
> $P$   is primary profile obtained from the transducer, which incorporates the $R$ and $W$.

These are defined in Section 2.1.

To specify a value, three characteristics should be incorporated. Thus, T.n.N represents the notation:

> T is the type of profile e.g. $R$,
> n is the parameter suffix e.g. q,
> $N$ is the number cut-offs (sampling lengths) e.g. 2.

So, if T.n.$N$ is $R_q2$, this means roughness profile, root mean square value taken over two sampling lengths.

For convenience, some values for $N$ have been adopted. These are listed in Table 3.1. If not otherwise indicated on a drawing, the following should be used to determine the cut-off wavelength (ISO 4288). Care has to be taken when using the letter $N$. Here, this is the number of sampling lengths and not the 'N' value of the surface, which is an old fashioned notation for surface roughness. See the end of this chapter.

**Table 3.1**  *Adopted values of N*

| Recommended cut-off ISO 4288 | | | | |
|---|---|---|---|---|
| Periodic profiles | Non-periodic profiles | | Cut-offs | Sampling length<br>Evaluation length |
| Spacing distance | $R_z$ (μm) | $R_a$ (μm) | $\lambda_c$ (mm) | $\lambda_c l$ (mm) |
| > 0.01 to 0.04 | to 0.1 | to 0.02 | 0.08 | 0.08/0.4 |
| > 0.04 to 0.13 | > 0.1 to 0.5 | > 0.02 to 0.1 | 0.25 | 0.25/11.25 |
| > 0.13 to 0.4 | > 0.5 to 10 | > 0.1 to 2 | 0.8 | 0.8/4 |
| > 0.4 to 1.3 | > 10 to 50 | > 2 to 10 | 2.5 | 2.5/12.5 |
| > 1.3 to 4 | > 50 | > 10 | 8 | 8/40 |

Procedures are as determined in ISO 4287/1, ISO 11 562 and other international standards.

## 3.2   Classification of parameters for the profile

These are amplitude, spacing and hybrid parameters as shown in Figure 3.1. Amplitude and spacing parameters are self-evident. Hybrid parameters are those obtained, usually indirectly, via the spacing and amplitude parameters e.g. slope.

The question of which parameter to use for surface roughness is a difficult one because of the lack of real information. The magnitude of the problem is large. For example, take a ground surface of 25mm diameter and 100mm long – a quite typical workpiece. The correlation length (the size of a typical hit by a grain) is about 10μm and the aspect ratio of the grinding mark is perhaps 50:1. So, the number of independent hits on the part is given by $N_1 = \pi . \dfrac{25 \times 100}{0.01 \times 50 \times 01} \sim 1M$ bits. Yet from the point of view of performance, all that is needed is to know whether the part is good or not! One bit of information. Yet there are millions to pick from as the calculation shows.

The number $N_i$ is the number of potential locations on the specimen where nominally independent height values can be located: it does not relate directly to good or bad bits of information. There is a difference. The starting point for surface failure could well be very small but, once initiated, develops as an avalanche e.g. in oil film breakdown. The starting point would have to be exceptional; e.g. an extra high peak or an extra deep valley but small in spatial size and number. On the other hand, to 'help' performance such as load-carrying capacity, many parts of the surface have to contribute. It is an average effect. A *failure* mechanism can be regarded therefore as being initiated by a *differential characteristic* of the surface, whereas beneficial effects are the result of *integrated characteristics*. Another distinction is that failures can be location-dependent by, for example, having a crack at the root of a turbine blade, whereas beneficial effects are only area-dependent.

The significance of the surface parameters will be dealt with elsewhere. It is brought up here to illustrate the problem of characterization.

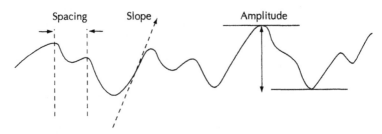

**Figure 3.1**   *Roughness parameter types*

Figure 3.1 shows the general grouping of surface parameters.

In general engineering in the past, that it has been the amplitude parameters that have attracted most attention rather than the spacings. The amplitude parameters were considered to be most relevant for functional behaviour e.g. in contact. On the other hand, spacings have been regarded as most useful in metallurgy and material science in general because of the need to examine structure in, for example, grain boundaries.

It is only recently that it has become obvious that both are needed. Also, closer scrutiny of the functional behaviour of surfaces in friction and wear situations have pointed to slopes and curvatures – the hybrid parameters. This muddle is now being corrected by modern instruments.

Another point to note is that, in the past, engineers have not been too concerned with areal properties such as the lay pattern, but they have attached great importance to profile parameters important in wear and contact, especially those features that can influence failure. It has become obvious only recently that the areal (3D) parameters determine flow behaviour. This has switched attention from purely profile parameters to areal (3D) parameters. Note that areal parameters are those obtained from an area of the surface rather than a line profile. Areal parameters are sometimes called three-dimensional, hence the term in brackets.

### 3.2.1   Amplitude parameters

In what follows, some of the available parameters will be described. The reason why there are so many with sometimes small differences is a consequence of their historical background.

### 3.2.1.1 The $R_a$ value

The $R_a$ value is the universally recognized and most used parameter of roughness. It is the arithmetic mean of the magnitude of the deviation of the profile from the mean line. $R_a$ is a relatively new symbol. It used to be called CLA (centre line average) in the UK and AA (arithmetic average) in the USA. The symbol $R_a$ is useful because it relates to roughness average in English and *Rauteife* in German, the two main camps in surface metrology.

How the $R_a$ value is worked out is illustrated in Figure 3.2 (a).

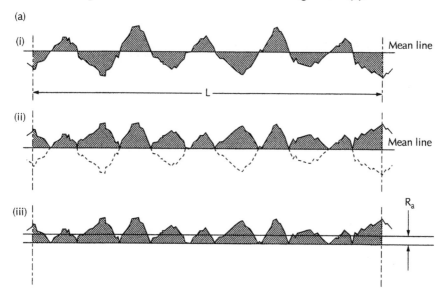

**Figure 3.2 (a)**  *$R_a$ parameter*

In terms of the chart, the mean line of the profile is first found by working out the average height of the profile over the length L – usually the sampling length. Once this is drawn in, it obviously bisects the profile so that an equal area lies above it as below it (Figure 3.2 (a)(i)). It is no good simply measuring the average deviation of the profile from the mean line: it has to be zero by definition. So, it is necessary to make all the negative deviations positive as in (ii) and then work out the average height of this rectified waveform from the mean line – which is now the base line. This average height is the $R_a$ value as shown in (iii).

$R_a$ (or any other parameter) alone is not sufficient by itself to describe the roughness on a wide range of surfaces. Figure 3.2 (b) (overleaf) shows some periodic waveforms all having the same $R_a$ values yet all have different peak-valley values and have different relationships between the $R_a$ and the peak-valley.

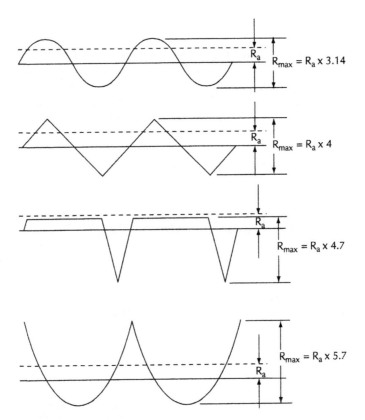

**Figure 3.2 (b)**   *Waveforms with equal $R_a$ values*

The formula usually adopted for the $R_a$ is given below.

$$R_a = \frac{1}{l_r} \int_0^{lr} |z(x)| dx \qquad (3.1)$$

where | | indicates that the sign is ignored and $z(x)$ is the profile measured from the mean line at position $x$.

### 3.2.1.2 The $R_q$ value

$R_q$ is the rms parameter corresponding to $R_a$ and, because it is more easily related to functional performance, is gradually superseding $R_a$.

$$R_q = \sqrt{\frac{1}{lr} \int_0^{lw} z(x)^2 \, dx} \qquad (3.2)$$

The waviness equivalent of $R_a$, is:

$$w_a = \frac{1}{l_w} \int_0^{lw} |z(x)| dx \tag{3.3}$$

where $z$ has the roughness filtered out.

$W_a$, $W_q$, $P_a$, and $P_q$ are the corresponding parameters of the waviness and primary profiles, respectively. (Note that $R_q$ is sometimes referred to as the **root mean square value**). Figure 3.3 shows that it is larger than $R_a$ – by a factor of between 10 and 25% depending on the process.

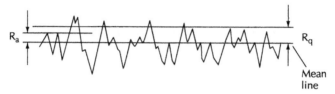

**Figure 3.3** *Root mean square*

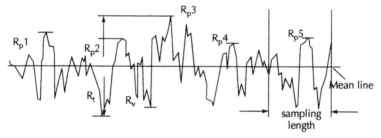

**Figure 3.4** $R_v$ $R_p$ $R_t$ $W_v$ $W_p t$ $P_v$ $P_p$ $P_t$

Figures 3.4–3.12 shows some typical peak and valley parameters.

$R_v$ is the maximum depth of the profile below the mean line within the sampling length.

$R_p$ is the maximum height of the profile above the mean line within the sampling length.

$R_t$ is the maximum peak to valley height of the profile in the assessment length.

$W_v$, $W_p t$, $P_v$, $P_p$ and $P_t$ are the corresponding parameters of the waviness and primary profiles, respectively.

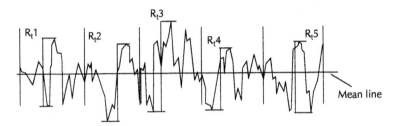

**Figure 3.5**  $R_{z}$, $R_{t}$, $R_{tm}$

$R_z = R_p + R_v$ and is the maximum peak to valley height of the profile within a sampling length.

$R_{ti}$ is the maximum peak to valley height of the profile in a sampling length $\equiv R_{zi}$.

$R_{tm}$ is the mean of all the $R_{ti}$ values obtained within the assessment length, where $N =$ the number of cut-offs then:

$$R_z \equiv R_{tm} = \frac{R_{t1} + R_{t2} + R_{t3} + R_{t4} + R_{t5}}{5} = \sum_{i=1}^{i=N} R_{ti} \tag{3.4}$$

Note that $R_{tm}$ is the equivalent of $R_z$.

Also, in the purely peak to valley parameters of the primary profile, the mean line is not needed in the calculation. It is always useful, however, to have the mean line pointing in the general direction of the surface by levelling.

$R_{tm}$ and $R_{ti}$ are non-ISO standards but are application-specific.

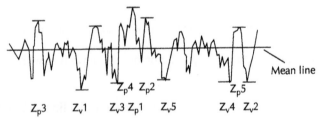

**Figure 3.6**  $R_z$(JIS) (ISO 10 point height parameter)

$R_z$(JIS), also known as the *ISO 10 point height parameter* in ISO 4287/1-1984, is measured on the roughness profile only and is numerically the average height difference between the five highest peaks and the five lowest valleys within the sampling length (JIS is the Japanese Industrial Standard).

$$R_z(\text{JIS}) = \frac{(Z_{p1}+Z_{p2}+Z_{p3}+Z_{p4}+Z_{p5})-(Z_{v1}+Z_{v2}+Z_{v3}+Z_{v4}+Z_{v5})}{5}$$

$$= \frac{1}{5}\left(\sum_{i=1}^{i=5}z_{pi} - \sum_{i=1}^{i=5}z_{vi}\right) \tag{3.5}$$

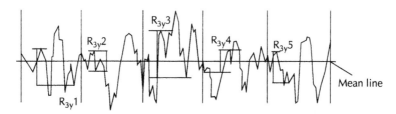

**Figure 3.7**  $R_{3y}$, $R_{3z}$

The deviation from the third highest peak to the third lowest valley in each sample length is found. $R_{3y}$ is then the largest of these values. The idea of this is to preclude rogue values of peaks and valleys. $R_{3z}$ is the vertical mean from the third highest peak to the third lowest valley in a sample length over the assessment length (DB N3 1007 (1983)).

where $N$ = number of cut-offs, then:

$$R_{32} = \frac{1}{N}\sum_{i=1}^{i=N}R_{3yi} = \frac{R_{3i1}+R_{3y2}+\dots R_{3yn}}{N} \tag{3.6}$$

The above parameters are non-ISO standards but are application-specific. They are never to be used unless their use is mentioned in the document.

### 3.2.3 Spacing parameters

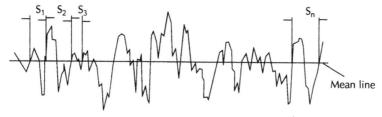

**Figure 3.8**  $RS_m$

$RS_m$ is the mean spacing between profile peaks at the mean line, measured within the sampling length. (A profile peak is the highest point of the profile between an upwards and downwards crossing of the mean line.)

If $n$ = number of peak spacings, then:

$$RS_m = \frac{1}{n} \sum_{1=1}^{i=n} S_i = \frac{S_1 + S_2 + S_3 \ldots + S_n}{n} \qquad (3.7)$$

$WS_m$, and $PS_m$ are the corresponding parameters from the waviness and primary profiles, respectively.

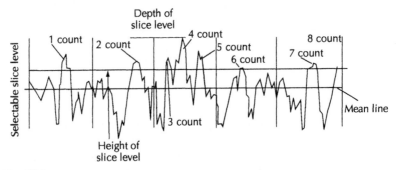

**Figure 3.9**   *HSC*

HSC, the **high spot count**, as shown in figure 3.9, is the number of complete profile peaks (within assessment length) projecting above the mean line, or a line parallel with the mean line. This line can be set at a selected depth below the highest peak or a selected distance above or below the mean line. It is therefore useful for specific applications.

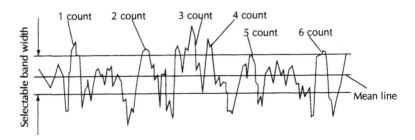

**Figure 3.10**   *P_c*

$P_c$, the **peak count**, is the number of local peaks that project through a selectable band centred about the mean line (see Figure 3.10). The count is determined only over the assessment length though the results are given in peaks per cm (or per inch). The peak count obtained from assessment lengths of less than 1cm (or 1 inch) is obtained using a multiplication factor. The parameter should, therefore, be measured over the greatest assessment length possible.

### 3.2.4  Hybrid parameters

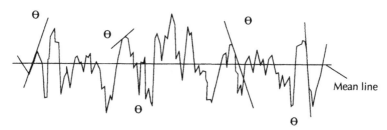

**Figure 3.11**  $R_{\Delta q}$

$R_{\Delta q}$ is the rms slope of the profile within the sampling length.

$$R_{\Delta q} = \sqrt{\frac{1}{lr} \int_0^{lr} \left( \theta(x) - \bar\theta \right)^2 dx} \tag{3.8}$$

$$\bar\theta = \frac{1}{lr} \int_0^{lr} \theta(x) dx \tag{3.9}$$

where $\theta$ is the slope of the profile at any given point. This is the mean absolute slope i.e. sign ignored. If the slope is not ignored, then $\bar\theta$ is obviously zero over the total profile.

### 3.2.4.1  $R_{\lambda q}$ Average wavelength, λq(rms)

This is a measure of the spacings between local peaks and valleys, taking into account their relative amplitudes and individual spatial frequencies. Being a hybrid parameter determined from both amplitude and spacing information, it is, for some applications, more useful than a parameter based solely on amplitude or spacing.

$$R_{\lambda q} = \frac{2\pi R_q}{\Delta q} \tag{3.10}$$

**Note**: the average wavelength is strictly a spacing parameter. It is the fact that it is evaluated by using height and spacing that allows it to be called 'hybrid'. Some acceptable and unambiguous definition of 'hybrid' is still needed.

$W_{\Delta q}$, $W_{\lambda q}$, $P_{\Delta q}$, and $P_{\lambda q}$, are the corresponding parameters of the waviness and primary profiles, respectively.

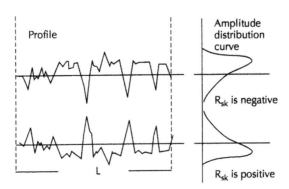

**Figure 3.12**   $R_{sk}, R_{ku}, W_{sk}, W_{ku}, P_{sk}, P_{ku}$

### 3.2.4.2   $R_{sk}$ – skewness

This is the measure of the symmetry of the profile about the mean line. It will distinguish between asymmetrical profiles of the same $R_a$ or $R_q$; $z$ is measured from the mean line.

$$R_{sk} = \frac{1}{R_q^3}\left[\frac{1}{lr}\int_0^{1r} z^3(x)dx\right] \qquad (3.11)$$

### 3.2.4.3 $R_{ku}$ – kurtosis

This is a measure of the sharpness of the surface profile.

$$R_{ku} = \frac{1}{R_q^4}\left[\frac{1}{lr}\int_0^{1r} z^4(x)dx\right] \qquad (3.12)$$

See later for an alternative form for $R_{sk}$ and $R_{ku}$.

$W_{sk}$, $W_{ku}$, and $P_{ku}$ are the corresponding parameters for the waviness and primary profiles respectively.

### 3.2.5   Curves

So far, the outcome of a characterization has been a number e.g. 10μm $R_a$. It is obvious, however, that more is needed in cases where performance needs quantifying. These need the mathematical relationship itself rather than just a final number from it. In some cases, a plot of values is needed. These have been called 'curves' although they are formulae evaluated at a number of points and a line fitted through the points.

A fixed value obtained for a parameter such as the $R_a$ value is possible because the formula is in 'closed form'. This means that there is no variable left with which to influence the number e.g. $\dfrac{1}{L}\displaystyle\int_0^L |z(x)|dx = 10\mu m$ means that all the z values from x = 0 to L have been taken into account. Nothing more needs to be done. There are, however, some formulae for parameters that are in 'open form': the result still contains a variable.

Taking the amplitude density distribution curve represented by $p(z)$ shown above alongside the skew figure, $p(z)$ is the probability that a z value occurs as the profile is scanned from left to right. If the z value is taken to be 5μm, $p(z)$ could be 0.1. If on the other hand z is taken to be 10μm $p(z)$ could be 0.05. The $p(z)$ value depends on the z value. Putting the number 0.1 down does not mean anything unless it is limited to the 5μm level at which it is being evaluated. If all the z values for the profile are taken into account, then the $p(z)$ values add up to unity. If the value of $p(z)$ is plotted as a function of z, a curve results as seen in Figure 3.13 on the right side of the profile.

**Figure 3.13**   *Probability density curve*

The skew value can be worked out from $p(z)$. It is a *number* describing some aspect (the symmetry) of the $p(z)$ curve and hence the profile.

So, the skew $R_{sk}$ is a number (closed form) obtained by operating on an open form curve $p(z)$.

Thus:

$$R_{sk} = \frac{1}{R_q^3}\int_{-\infty}^{\infty} z^3 p(z)dz \qquad \equiv \frac{1}{R_q^3}\frac{1}{l_r}\int_0^{l_r} z^3(x)dx \tag{3.13}$$

$$\text{(i)} \qquad\qquad\qquad \text{(ii)}$$

The skew value can be found (i) via a curve or (ii) in a closed form directly. In this case, the variable is z for (i) and x for (ii). The form (i) for $R_{sk}$, $R_{ku}$ and $R_q$ is referred to as the surface *moments*. Kurtosis, which describes 'spikiness' of the profile, can be evaluated either way by using (i) or (ii).

$$R_{ku} = \frac{1}{R_q^4} \int_{-\infty}^{\infty} z^4 p(z)dz \qquad \equiv \frac{1}{R_q^4} \frac{1}{l_r} \int_0^{l_r} z^4(x)dx \qquad (3.14)$$

$$\text{(i)} \qquad\qquad\qquad \text{(ii)}$$

Obviously, from $p(z)$ or $z(x)$, the value obtained for $R_{sk}$ is the same. Also, the value obtained for $R_{ku}$ by the two methods is the same. The difference is that $p(z)$ builds up a picture of the characteristics of the profile as $z$ changes, whereas a succession of $z$ values at $x$ positions tells very little. A 'function', described here as a 'curve', e.g. $p(z)$ is in effect another way of picturing the profile. The information in the surface is presented in a way that is different from the profile and, sometimes as will be seen later, clearer.

The ability of curves rather than single numbers to show up different characteristics of the surface seems to be indisputable, yet they are not usually given as making up the formal definitions for $R_{sk}$ or $R_{ku}$ in the ISO documents 4387:1997! This omission is a serious fault in ISO documentation to date.

As before, in order to rationalize the measurement procedure, the same parameters (and curves) are designated for waviness and the primary profile as they are for roughness. Thus $P_{sk}$, $R_{sk}$, $W_{sk}$ and $P_{ku}$, $R_{ku}$ and $W_{ku}$ are the families of parameters for the surface profile.

There are similar families that embody the 'curve' itself rather than a value obtained from it. One such example is the **material ratio curve**.

### 3.2.6   Material ratio curve

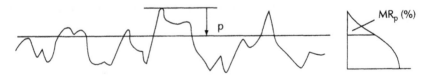

**Figure 3.14**   *Material ratio curve*

This is a measure of the material to air ratio expressed as a percentage and has the depth p below the highest peak included. Thus, a particular value of the curve shown in Figure 3.14 is the percentage of material to air at depth p.

The material ratio curve is unique because it was the first 'functional parameter'. It was devised by Dr Abbott [3.1]. His idea was to describe how useful a surface might be when used as a bearing. He devised this ratio of material to air starting from where he considered the first contact would be i.e. at the highest peak and showing how it changed with depth (Figure 3.15). The problem with this definition is that it is difficult to know where the first contact is. The highest peak within one assessment profile is not necessarily the overall highest peak. For this reason, the ordinate axis is sometimes put in terms of the $R_q$ value. The actual curve is the same. The top of the $MR$ curve is in a different place. This is not important in the functional sense.

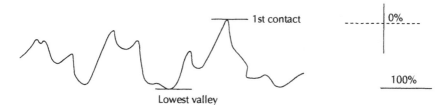

**Figure 3.15**  *Material ratio curve*

During its years of use, the curve has had a number of names. These are:

Abbott-Firestone curve  
Abbott-curve  
Bearing ratio curve          These are all the same.  
Bearing area curve  
Material ratio curve

The actual calculation is straightforward. The value $R_{MR}(Tp\%)$ is the ratio in Figure 3.14 of the material lengths $b_1 + b_2$ etc. expressed as a percentage of the total assessment l.

The enduring presence and use of the material ratio curve over the years led to attempts to utilize the shape of the $MR$ curve. One first attempt was due to Pesante [3.2] and Ehrenreich [3.3] who advocated differentiating the curve. They did this to overcome one of the problems of the material ratio curve, which is that it is somewhat insensitive to the value of $z$. What they were in fact doing was to plot the amplitude density function. The relationship between them is:

$$MR(h) = \left(1 - \sqrt{\frac{2}{\pi}} \int_{-\infty}^{h} p(z)dz\right) = 1 - erf(h) \tag{3.15}$$

More recently, the $MR$ curve has been the basis of a German proposal to measure stratified surfaces i.e. surfaces having a profile made up from two or more processes. Also, the method is proposed for use in highly stressed surfaces in ISO 4287 1996 and ISO 13565.

Some key definitions are:

(a) Roughness core profile, which excludes the high peaks and valleys (Figure 3.16).

(b)  Core roughness depth $R_k$ – the depth of the roughness core profile.

(c)  Material portion $MR1$ – The level in per cent determined for the intersection line that separates the high peaks from the roughness for the profile, $MR2$ is the corresponding level for the deepest valleys.

(d)  Reduced peak height $R_{pk}$ – average height of the highest peaks above the roughness core profile and $R_{vk}$ the corresponding average valley depth.

There are special filtering methods, which should be used in a material ratio curve analysis. These will be indicated later.

It should be emphasized that the method indicated below is not very scientific. For example, the material ratio has to be plotted with a linear abscissa, otherwise the 'S' shape needed for parameter evaluation might not occur.

If the $MR$ of a random surface e.g. in grinding is plotted, a straight line, not an S curve, results when $MR$ is plotted against a probability abscissa – which is the correct way.

### 3.2.6.1   Determination of Parameters

1.    *Roughness profile* – the roughness profile used for determining the parameters that are the subject of this part of ISO 13565 are calculated according to ISO 13565-1.

2.    *Calculating the parameters $R_k$, $MR1$, $MR2$* – the equivalent straight line intersects the abscissae $MR = 0\%$ and $MR = 100\%$ . From these points, two lines are plotted to the x-axis, which determine the roughness core profile by separating the protruding peaks and valleys. The vertical distance between these intersection lines is the core roughness depth $R_k$. Their intersections with the material ratio curve defines the material ratios $MR1$ and $MR2$.

3.    *Calculating the equivalent straight line* – the equivalent straight line is calculated for the central region of the material ratio curve, which includes 40% of the measured profile points. This 'central region' lies where the secant of the material ratio curve over 40% of the material ratio shows the smallest gradient (see Figures 3.16–3.21). This is determined by moving the secant line for $\Delta MR = 40\%$ along the material ratio curve, starting at the $MR = 0\%$ position as in Figure 3.17. The secant line for $\Delta MR = 40\%$, which has the smallest gradient, establishes the 'central region' of the material ratio curve for the equivalence calculation. If there are multiple regions that have equivalent minimum gradient, then the one region that is first encountered is the region of choice. A straight line is then calculated for this 'central region', which gives the least square deviation in the direction of the profile ordinates.

**Note 1**: where the 40% value originated is not clear.

**Note 2**: to ensure the validity of the material ratio curve, the class widths of ordinates of the roughness profile should be selected to be small enough for at least 10 classes to fall within the 'central region'. With surfaces having very small rough-ness or having an almost ideal geometrical plateau, such a fine classification may no longer be meaningful because of the limited resolution of the measuring sys-tem. In this case, the number of classes used in the calculation of the equivalent straight line may be stated in the test results.

4.    *Calculating the parameters $R_{pk}$ and $R_{vk}$* – the areas above and below the region of the material ratio curve that delimit the core roughness $R_k$ are shown hatched in Figure 3.16 and 3.17. These correspond to the cross-sectional area of the profile peaks and valleys that protrude out of the roughness core profile.

The parameters $R_{pk}$ and $R_{vk}$ are each calculated as the height of the right-angle tri-angle, which is constructed to have the same area as the 'peak area A1', has *MR1* as its base, and that corresponding to the 'valley area A2' has 100% – *MR2* as its base.

**Note**: The parameters according to this part of ISO 13565 should only be calculated if the material ratio curve is 'S' shaped as shown in Figures 3.20 and 3.21 and thus has only one single point of inflection. Experience has shown that this is always the case for lapped, ground or honed surfaces.

In the parameter evaluations, the procedures are arbitrary. The 40% base for $R_k$ is by agreement. It does not have a scientific basis yet it has been found to be useful. According to car engine makers, the curve of *MR* and parameters evaluated from it are functionally important in the case of plateau honing for cylinder liners.

Thus:

$R_{pk}$ is the amount worn down quickly when the engine runs,

$R_k$ determines the life of the cylinder,

$R_{vk}$ indicates the oil retention and debris capability of the surface.

Plateau honed surfaces bring to a head some of the recent problems associated with surface metrology. The biggest problem is that the surface and hence the profile is made from two distinct processes. One is rough honing and the other is fine honing or grinding. Such a profile is shown in Figure 3.22 (a). It is character-ized by having a fine random surface interspersed with very deep valleys. The $R_k$ parameters lend themselves readily to describing this sort of 'striated' or 'strati-fied' surface.

The two components can be dealt with together by using a long sampling length, say 2.5mm, which reduces the impact of the deep valleys. In terms of filter-ing, the weighting function of the filter should be long compared with the width of the valleys. This has the effect that the mean line does not sink into the valleys as seen in Figure 3.22 (a). The longer the cut-off of the sampling length, the less the dip (Figure 3.22 (b)). Any dip of the mean line reduces the effective depth of the valleys, which can give a misleadingly low estimation of the oil capacity of the surface. If the filter is not phase corrected, the effect is worse (Figure 3.22 (a)). Not only are the valleys apparently shallower, but the load-carrying part of the surface is distorted by giving the impression that there are high peaks (Figure 3.22 (b)).

The effect of not using a valley cut-off can be quantified [3.3]. Basically, the cut-off should be as long as possible e.g. 2.5mm to prevent the mean line dropping into the valleys (Figure 3.22).

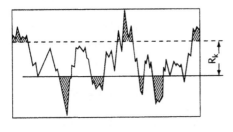

**Figure 3.16**   *Layering of profile – core*

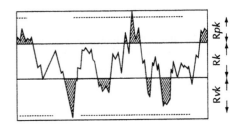

**Figure 3.17**   *Identification of 40%*

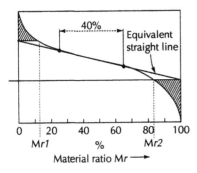

**Figure 3.18**   *Identification of Mr1 and Mr2*

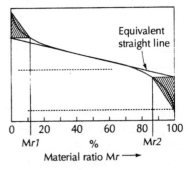

**Figure 3.19**   *Layering of profile $R_k$, $R_{pk}$, $R_{vk}$*

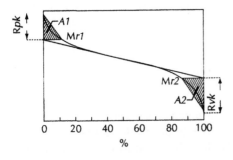

**Figure 3.20**   *Identification of areas A1, A2*

**Figure 3.21**   *Material ratio curve*

**Figures 3.16–3.21**   *Choice of filter and sampling lengths*

(a) 0.8mm

(b) 2.5mm

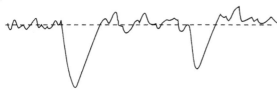

**Figure 3.22**   *Choice of sampling lengths*

(a) Profile

(b) Modified profile

**Figure 3.23**   *Standard 2CR filter*

## $R_k$ Filtering

This filtering technique is in accordance with ISO 13565 pt. 1 and DIN 4776.

### 3.2.6.2 Filtering process to determine the roughness profile

The filtering process is carried out in several stages, given the modified profiles (Figure 3.23).

The first mean line is determined by a preliminary filtering of the primary profile with the phase correct filter in accordance with ISO 11562 using a cut-off wavelength $\lambda_c$ in accordance with Clause 7 and corresponding measuring conditions in accordance with Table 1 of ISO 3274:1996 (Table 3.2 in this book). All valley portions that lie below this mean line are removed. In these places, the primary profile is replaced by the curve of the mean line.

The same filter is used again on this profile with the valleys suppressed. The second mean line thus obtained is the reference line relative to which the assessment of profile parameters is performed. This reference line is transferred to the original primary profile and the roughness profile according to this part of ISO 13565 is obtained from the difference between the primary profile and the reference line. [1.2]

The selected cut-off wavelength is $\lambda_c = 0.8$mm. In justified exceptional cases, $\lambda_c = 2.5$mm may be selected and this should be stated in the specification and test results.

**Table 3.2**   *Relationship between the cut-off wavelength $\lambda_c$ and the evaluation length ln*

| $\lambda_c$ | ln |
|:---:|:---:|
| 0.8 | 4 |
| 2.5 | 12.5 |

Both the $R_k$ filter and the long cut-off are compromises. This is inevitable because there is more than one process present in the profile. Attempting to embody both with one procedure is bound to produce errors.

*Scope*
ISO 13565 describes a filtering method for use with surfaces that have deep valleys below a more finely finished plateau with a relatively small amount of waviness. The reference line resulting from filtering according to ISO 11562 for such surfaces is undesirably influenced by the presence of the valleys. The filtering approach described in Figure 3.24 suppresses the valley influence on the reference line such that a more satisfactory reference line is generated. However, a longer cut-off should always be tried.

*Normative references*
The following standards contain part of ISO 13565. At the time of publication, the editions indicated are valid.

ISO 3274:1996, Geometrical Product Specifications (GPS) – surface texture: profile method nominal characteristics of contact (stylus) instruments.

ISO 4287:1997, Geometrical Product Specifications (GPS) – surface texture: profile method – terms, definitions and surface texture parameters.

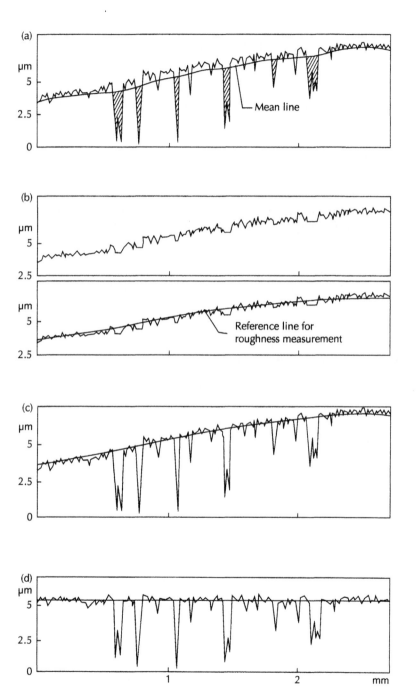

**Figure 3.24**   *$R_k$ filtering (a) unfiltered primary profile (valleys shown hatched) (b) unfiltered primary profile after suppression of valleys (c) position of the reference line in the primary profile (d) roughness profile in accordance with this standard*

ISO 11562:1996, Geometrical Product Specifications (GPS) – surface texture: profile method.

*Definitions*
For the purposes of this part of ISO 13565, the definitions given in ISO 3274 and ISO 4287 apply.

## ISO 13565-1:1996(E)

*Reference guide*
To measure profiles in accordance with this part of ISO 13565, a measuring system that incorporates an external reference is recommended. In case of arbitration, the use of such a system is obligatory. Instruments with skids should not be used.

*Traversing direction*
The traversing direction should be perpendicular to the direction of lay unless otherwise indicated.

Another point that causes confusion is concerned with the material ratio obtained from the profile and that obtained over an area. In fact, these are equal. The profile *MR* squared is not the areal value. See Figure 3.25 (a), which shows a series of blocks of profiles. This can be moved about to look like Figure 3.25 (b), which is still the same *MR* value yet it is the complete areal view from whichever direction it is taken.

If a large number of tracks are made on Figure 3.25, the material ratio is always the same. It is not valid to take just one reading on the lower surfaces. This equivalence is one reason why it is valid to base the use of the various functional parameters on the material ratio.

(a)

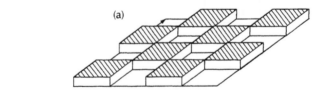

(b)

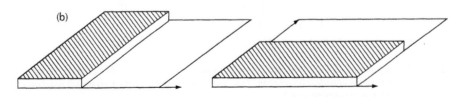

**Figure 3.25** *Areal material ratio*

### 3.2.7   Summary of simple parameters

Looking at the standards reveals a bewildering mix of parameters. The variety and number of parameters is not because of their individual usefulness. For the majority of parameters, their presence is a result of the historical development of the subject. This is unfortunate because it implies that each parameter is unique in its usefulness. This is not true. For example, the marginal differences in value between $R_a$ and $R_q$ are not significant functionally: either will do.

One of the reasons for the number of parameters is that two schools of thought developed (in much the same way as in reference lines). The UK and USA considered average parameters such as $R_a$ to be most useful and also appreciated the ease of measurement off a chart. Germany and the then USSR standardized on peak parameters $R_t$ etc. using the argument that the peaks were important in functional situations. Also, measuring peaks and valleys from a chart fitted in with the existing practice of measuring form from interference fringes. Both average and peak types of parameter are embedded in national and international standards.

Unfortunately, peak-valley type measurements are not 'stable'. They are described as 'divergent'. In other words, the longer the inspector looks for the maximum peak-valley, the bigger the value will be. In order to correct for this, the peak type parameters have had to be modified to make the values relatively stable. The modification in most cases is simple averaging, so that instead of $R_t$ the maximum peak to valley $R_{tm}$ was introduced. This was and is an average of a number of peak-valley values across the chart. To add to the confusion, some parameters have more than one name. For example, $R_a$ was originally known in the UK as CLA (*centre line average*) and in the USA as AA (*arithmetic average*). This early AA was in fact the root mean square roughness (now called $R_q$). It was divided by 1.11 to give the correct value of AA for a sine wave. Unfortunately, it gave the wrong answer for random type surfaces such as grinding, which requires a factor of 1.25.

All these circumstances have given rise to what has been termed the **parameter rash** [3.4]. There is absolutely no evidence that parameters of the same type e.g. peak amplitude parameters give different functional results within the type.

As if this is not confusing enough, some countries try to classify the parameters in terms of size. One typical classification is shown in Figure 3.26. This is called the N system. Another system uses a set of del symbols ∇∇∇ etc. The problem here is that standards documents can become cluttered up with what are in effect 'in-house' characterization methods used by large companies. Table 3.3 shows the hopeless confusion for some countries. High N means rough for some countries yet for others it means smooth. Unfortunately, the N numbers are still used! Great care has to be taken to make sure which N scale is being used. Also remember that the letter N is used for the numbering of sampling lengths in the assessment length.

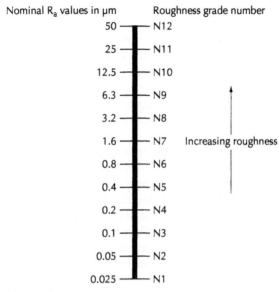

**Figure 3.26** *Roughness grades*

**Table 3.3** *N values*

| CLASS | UK | USA | GERMANY | USSR | JAPAN | ISO |
|---|---|---|---|---|---|---|
| 1 | 1 | 0.25<br>0.5<br>1.0 | 400/630 | 200 | 200/280 | 200 (R) |
| 2 | 2 | 2 | 160/250 | 125 | 100/140 | 125 |
| 3 | 4 | 4 | 63/100 | 63 | 50/70 | 63 |
| 4 | 8 | 8 | 25/40 | 40 | 25/35 | 40 |
| 5 | 16 | 16 | 16 | 6.3 | 18 | 6.3 |
| 6 | 32 | 32 | 6.3 | 3.2 | 12 | 3.2 ($R_a$) |
| 7 | 63 | 63 | 4.0 | 1.6 | 6 | 1.6 |
| 8 | 125 | 125 | 2.5 | 0.8 | 3 | 0.8 |
| 9 | 250 | 250 | 1.6 | 0.4 | 1.5 | 0.4 |
| 10 | 500 | 500 | 1 | 0.2 | 0.8 | 0.2 |
| 11 | 1000 | 1000 | 0.63 | 0.1 | 0.4 | 0.1 |
| 12 | – | – | 0.4/0.25 | 0.05 | 0.2 | 0.05 |
| 13 | – | – | 0.16/0.1 | 0.12 | 0.1 | 0.025 |
| 14 | – | – | 0.06/0.04 | 0.06 | – | 0.012 |
| Unit<br>Standard | BS1134<br>1950 | B46<br>1955 | 4763 DIN<br>1954 | GOST2780<br>1951 | JIS<br>1955 | 1953 |

**Table 3.4**   *Finish nomenclature by process*

| Turning | T | Boring | B |
|---|---|---|---|
| Diamond turning | DT | Reaming | R |
| Grinding | G | Milling | M |
| Honing | H | Planing | P |
| Lapping | L | Scraping | S |
| Polishing | Po | Broaching | Br |

Note: Obviously this shorthand is different in different languages.

In some documents, the finish is referred to by process. Table 3.4 gives a few typical examples.

## 3.3 Random process analysis

### 3.3.1 General

The collection of parameters listed above as amplitude spacing and hybrid were established mainly on an ad hoc basis with little regard to forming a coherent set. The amplitude distribution and material ratio, however, fit into the class called **random processes**. In its simplest form, the process functions are the **amplitude distribution** (or the **materials ratio**; which is its integral). The amplitude distribution has shape, position and scale and can be the basis for the amplitude parameters given above. Likewise, there is a similar function for the basis of the horizontal information. This can take two forms. One is the autocorrelation function. The other is the power spectral density.

### 3.3.2 Explanation

A profile is quite complicated. Any one point can be made up of random and/or deterministic information. It is very difficult to identify the components of manufacture simply from the chart – there are too many unknowns. Random process analysis is a way of reducing the unknown factors, thereby allowing the basic mechanisms of the surface generation to be revealed. How do the amplitude distribution (or $MR(\%)$) and autocorrelation function (or power spectrum) remove irrelevant information? The answer is that both remove positional information from the profile. The amplitude distribution looks at what is left in terms of the heights of the profile and the autocorrelation looks at spacings (but not position).

Figure 3.27 (a) shows that two profiles on the same surface i.e. (i) and (ii) produce the same amplitude distribution (c). It does not matter which order the data is in. For example, $P_1$ can be examined before or after $P_2$, it makes no difference to the amplitude distribution (c).

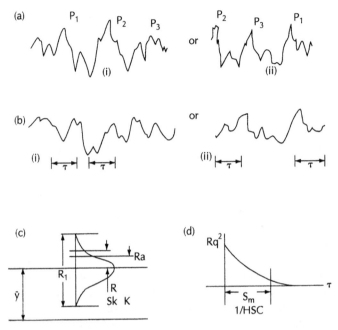

**Figure 3.27**   *Random process analysis*

In Figure 3.27 (b), the autocorrelation looks at the average value of the product of the profile 'spaced apart' by $\tau$. The position of each pair of ordinates separated by $\tau$ is irrelevant. What is shown in (d) has all the spacing information.

It is therefore true to say that one of the biggest problems, that of random positional effects, has been suppressed. Therefore neither the amplitude distribution nor the autocorrelation function can be as variable as the profile waveform simply because one random element has been removed. Consequently, the amplitude distribution and the autocorrelation function are 'stable'; they have more or less constant shapes wherever on the surface the profile has been taken. It makes sense therefore to use these two 'curves' as a basis for characterization.

Typical parameters estimating the shape and size are shown in Figure 3.27 (c) for amplitude distribution and Figure 3.27 (d) are for the autocorrelation. Figure 3.27 (d) shows some spacing parameters relative to the autocorrelation function of the profile.

Figure 3.28 shows some examples of these random process analysis functions for two typical processes: grinding and turning. Take (a) for grinding first. The profile has its amplitude distribution to the right, representing height information. The material ratio is another way of expressing the amplitude distribution and is to some extent more useful, as has been shown. Notice the Gaussian (bell) shaped curve. The autocorrelation is below the profile. It shows that for a relatively small value of the ordinate separation $\tau_l$ the correlation is zero. This means that the ordinates are not correlated for spacings bigger than $\tau_l$. This is to be expected because the grain size of the wheel producing the profile is not larger than $\tau_l$ in size.

For turning, the situation is different. The amplitude distribution is not Gaussian because the profile is not random. The autocorrelation function reflects the feed mark wavelength $\lambda_f$ as is to be expected because turning is essentially deterministic i.e. can be predicted knowing the process parameters. For processes such as single point cutting, it is visually more rewarding to use the power spectrum than it is to use the autocorrelation function. Notice the presence of harmonics as well as the feed component in the spectrum because the tool mark is not sinusoidal.

The autocorrelation function is relatively easy to compute. It is similar to the process of convolution. Instead of a shaped window or weighting function representing the filter response, the window is the function itself. The output is the average multiplication of the function with itself shifted by $\tau$. This is shown in Figure 3.29.

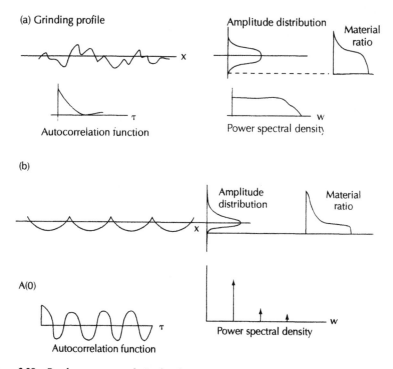

**Figure 3.28** *Random process analysis of surfaces*

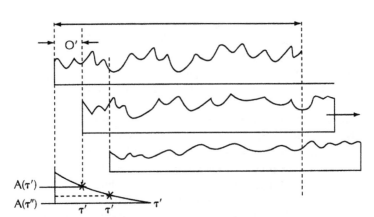

**Figure 3.29**  *Mechanism of autocorrelation function*

The only difference between convolution and correlation is that in the latter the average value of the products is taken, whereas in convolution it is not.

The operation on the chart to get the autocorrelation function is simple. After making sure that the mean value has been taken from the profile, the following procedure is followed.

(1) Multiply each ordinate of the profile by itself across the whole profile and add together all products. Divide by the number of ordinates multiplied. This is plotted on the small graph at the bottom of the Figure 3.29. If the ordinates are $z_1$ $z_2$ ...$z_n$:

$$A(0) = \frac{1}{N} \sum_{i=1}^{N} z_i^2 \tag{3.16}$$

(2) Shift the profile relative to itself (as the second profile in Figure 3.29) by an amount $\tau'$. Multiply each ordinate on the top profile by the ordinate (on the lower shifted profile) directly below it. Add the products and divide by the number of products. This gives $A(\tau')$, which is plotted at $\tau'$ on the small graph at the bottom ($\tau'$ is a number of ordinate spacings of profile shift).

$$A(\tau') = \frac{1}{N - \tau'} \sum_{i=1}^{N} z_i \cdot z_{i+\tau'} \tag{3.17}$$

(3) Repeat the process of shifting the second profile, multiplying vertical corresponding ordinates and take the average giving $A(\tau'')$, $A(\tau''')$ etc.

The bottom graph is called the **autocorrelation function** and contains all the spacing information. Correctly, this graph is sometimes called the **autocovariance function** because average values are removed, but usage favours the 'autocorrelation function'. Also, $A(\tau)$ is sometimes normalized by dividing by $R_q^2$ (the profile variance i.e. the rms$^2$ value).

The actual curve of $A(\tau)$ is the autocorrelation function. This contains the spacing information. In analogue form:

$$A(\tau) = \frac{1}{L - \tau} \int_{0}^{L-\tau} z(x).z(x + \tau)dx \qquad (3.18)$$

where the integral sign takes over from the summation in $A(\tau')$ above.

The power spectrum is the Fourier transform of the autocorrelation function in the same way as the profile is analysed and transformed to frequency in Figure 3.30 (a). The basic behaviour of the autocorrelation function is:

1.  It is always maximum at the origin $A(0)$.
2.  A random profile $A(\tau)$ decays to zero at a finite value of $\tau$.
3.  It does not matter in which direction the shift is made.
4.  A periodic profile $A(\tau)$ is periodic of the same wavelength – it does not decay.
5.  Additive components of the profile also add in correlation .

Some examples are shown in Figures 3.30 (a) and (b).

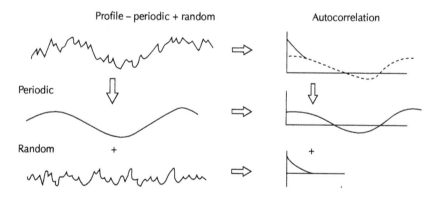

**Figure 3.30 (a)**   *Examples of use of autocorrelation function: small random component*

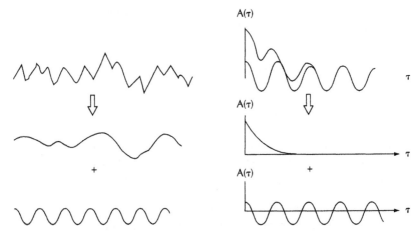

**Figure 3.30 (b)**   *Examples of use of autocorrelation function: large random component*

Example 1 could be grinding with chatter.
Example 2 could be of turning with microfracture of the material.

In summary, the autocorrelation can identify randomness from periodic signals. Consider a profile made up of a periodic and a random component as shown:

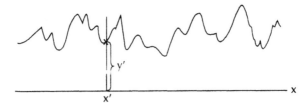

**Figure 3.31**   *Profile*

At every value of $x$, $y(x) = r(x) + p(x)$,     $r$ is a random component,
                                            $p$ is a periodic component.

It does not matter what the value of $x$ is.

**Figure 3.32**   *Autocorrelation A(τ )*

On the other hand, if instead of the profile the autocorrelation function is examined, it is seen that whilst at large values of $\tau$ this is no longer true i.e.

$$A(\tau) = A_p(\tau)$$

Simply extrapolating $A_p(\tau)$ back to the origin from the case where $\tau$ is large gives the variances of both $p(x)$ and $r(x)$ i.e. $R_{qp}^2$ and $R_{qr}^2$.

The extraordinary clarity of the autocorrelation function and the power spectrum is because each point on the function is the sum of a large number of multiplications and additions! Also, random phase effects are ignored, thereby exposing the random amplitude effects in the profile. This is a consequence of the suppression of position.

### 3.3.3   Conclusions

So far, a number of parameters have been described and some functions or 'curves' e.g. material ratio, autocorrelation etc have been described. There seem to be too many. This should be qualified to too many **redundant parameters**. In practice, the integrated and average parameters are best because they are measurable and stable.

Everything that has been said about surface parameters so far has been about a single surface, not two or more, which is likely to be the situation in practice. Also, profiles have been investigated and not the overall 'areal' surface, which is far more descriptive. From what has been said, it is obvious that many of the parameters could be omitted from a working set. For example, it is meaningless to put down $R_a$, $R_q$ or $R_{tm}$ as parameters and leave off spacing information. What is important is classifying surfaces into fundamental types of which amplitude is one, spacing is another and so on.

Another way of looking at the classification problem is to try to classify the type of statistics involved in any given application. This will be considered shortly. It should be pointed out that the simple parameters such as $R_a$ or $R_z$ are very suitable for use in process control. Their values and variability just have to reflect the health of the process.

If the interest is in the machine tool and its characteristics, then the more sophisticated 'functions' rather than simple parameters should be used. Here, the autocorrelation function and the power spectrum are relevant. There is evidence that they are powerful in machine tool diagnostics: they can reveal small changes in the path of the tool, for example, which would not be detectable by the usual parameters.

So, there are two ways forward. One is to extend the characterization to include areal (3D) information and the other is to consider the functional importance of surfaces. In Section 3.4, areal information and statistical types will be considered.

## 3.4 Areal (3D) assessment

### 3.4.1 General

Figure 3.33 attempts to put in perspective the role of areal measurement with respect to the profile. Note that the term *(3D) roughness* is wrong. The correct term is **areal (2D)** as compared with **profile (1D)**. However, the 3D is put in brackets for clarity. A profile has one demand; axis 'x' and one dependent measurement from it, 'z'. Areal measurement has two demand axes 'x' and 'y' and one dependent measurement from it, 'z'. Areal information can be useful in monitoring the condition of the machine tool. For example, errors in the path of the tool due to chatter can be assessed. Process information can be investigated via the profile. Built-up edge in turning or dressing requirement as in grinding can be extracted from profile information using autocorrelation or the power spectrum.

Similarly in terms of the 'function', the areal information plays a very significant role whenever fluid flow is involved parallel to the surface or surfaces.

Areal information is also important in contact situations because the longest wavelength in either lateral direction determines the positions of initial contact. Usually, this initial point of contact elastically deforms under load. The extent of this affects the actual contact. It is in this regime where detailed local information, which can be obtained from the profile, determines performance. In the case of contact energy flow normal to the surface such as electrical and thermal energy, it is greatly influenced by the local surface geometry such as peak curvature, whereas the areal geometry determines the distribution of contacts in space.

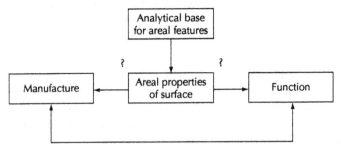

**Figure 3.33**   *Basis or areal characterization*

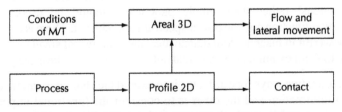

**Figure 3.34**   *Importance of profile (2D) characterization with respect to areal (3D)*

### 3.4.1.1    Types of surface information

Rather than getting confused by the large number of parameters, it is better to lump the parameters into 'types'. One attempt is shown in Figure 3.35. For simplicity, only one surface is considered. A typology is given in the figure.

1. Simple average e.g. $R_a$ of profile

2. Simple extreme $R_p$ $R_z$ $R_{TM}$ of profile

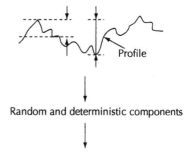

Profile

3. Areal (3D) information (average statistics)

Random and deterministic components

4. Areal information (change in statistics)

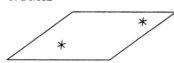

Wigner distribution ambiguity function
Wavelet function
Space frequency functions

5. Defects

**Figure 3.35**    *Types of surface information*

Consider the problem of characterizing the areal (3D) surface. Two difficult examples are given below.

**(i) Plateau honing**

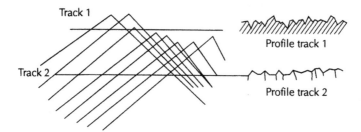

**Figure 3.36**    *Hone reversal in plateau honing*

A conventional track across the surface yields very little, as can be seen on the right of the illustration. Ideally, the instrument stylus must mimic the tool path. Deviations from the tool path detected by the stylus point to errors in the machine.

At the position where the honing tool changes direction, the characterization is most difficult.

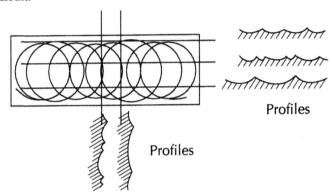

Profiles

Profiles

**Figure 3.37**   *Complex lay pattern milling*

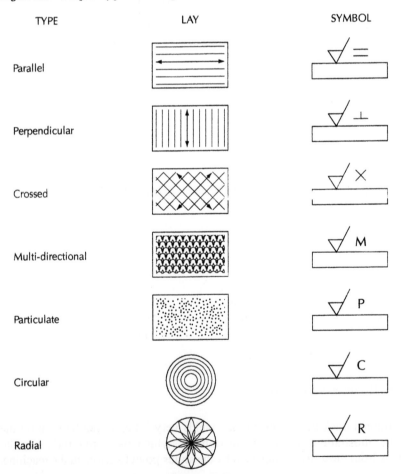

| TYPE | LAY | SYMBOL |
|------|-----|--------|
| Parallel | | = |
| Perpendicular | | ⊥ |
| Crossed | | × |
| Multi-directional | | M |
| Particulate | | P |
| Circular | | C |
| Radial | | R |

**Figure 3.38**   *Lay and symbols used on drawings [1.2]*

A very common process, milling, is very difficult to characterize using profiles yet the lay pattern is actually quite simple, being a combination of a rotation with a linear movement.

Figures 3.36 and 3.37 illustrate the complexity of any characterization. The path of the machining marks even in well-known processes make a straightforward description difficult. There has been some attempt to classify the lay on the surface, if only pictorially, as can be seen in Figure 3.38.

### 3.4.2 Relationship between profile and areal parameters

It has been common practice to estimate areal parameter values from the corresponding profile values. Parameters such as peak density have long been estimated this way. This practice is not only misleading, it is also responsible for the delay in realizing the importance of areal data in its own right.

The density of summits is in general not equal to the square of the density of peaks because some so-called peaks in the profile are **cols (saddle points)** (summit is the name given to an areal peak).

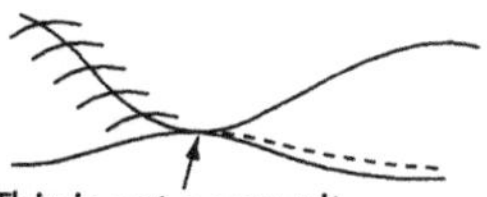

This is not a summit,
it is just a saddle point

**Figure 3.39**   *Problems with summits*

The number of (peaks/unit length)$^2$ is about 20% higher than the number of summits/unit area. Material ratio taken from a profile should not be squared to get areal information. In most cases of anisotropy and isotropy, the profile material ratio is the same as the areal material ratio (Figure 3.40).

The fact is that cross terms cannot be anticipated. Squaring profile parameters can only be used to get some idea of areal parameter values. Cross terms cannot be anticipated from profile graphs unless a number are taken in specified orientations. Even then, assumptions have to be made.

Some parameters need more than just a number or set of numbers to represent them. One good example is directionality, as shown in Figure 3.42. This needs to be coupled to another surface in order to make sense. This parameter is more like a vector than a scalar quantity because the direction has to be specified. If this sort of surface is mounted in the wrong direction in a bearing, the result can be disastrous. All the numerical values $R_a$, $R_q$ etc. are the same in (a) and (b) but the direction of

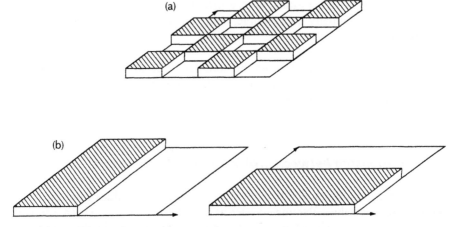

**Figure 3.40**   *Problems with material ratio*

motion relative to the loading edge is different. Notice that this problem occurs in lateral motion, not movement perpendicular to the surfaces. Figure 3.41 shows how in the future the system of surfaces might have to be simulated for every application.

Discussion of directionality is important because it shows exactly what is wrong with surface metrology today. Too much emphasis is being placed on single surfaces and not enough for the system as a whole, which usually consists of two surfaces.

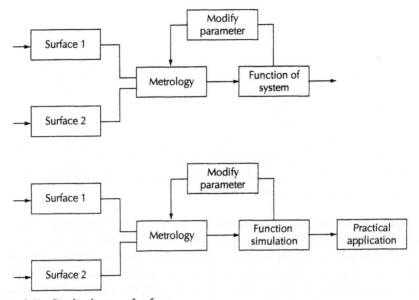

**Figure 3.41**   *Simulated system of surfaces*

The choice is to *modify* the geometrical parameter with what is perceived to be functional information or to modify geometrical parameters by simulation.

This shows the futility of describing one surface without taking into account the way in which the system i.e. the bearing is operated. Pictures of lay as in Figure 3.38 might be useful cosmetically in cases where only one surface is being used, but is not much use where two surfaces are concerned unless qualified with dynamic information.

So, a surface classification for function cannot be separated from the system i.e. slope is temporal and not spatial.

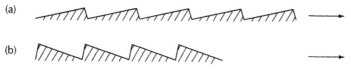

**Figure 3.42**   *Directionality is a factor of importance*

The performance of the surface as part of a bearing is completely different for (a) and (b) despite being the same surface. In other words, the surface cannot be separated from the system.

### 3.4.3   Comments on areal parameters [B2]

Despite its slow start, there have been some attempts to parameterize the whole surface. This effort has been led by Stout [3.5] who has been active in European Standards committees and has produced some useful, if interim, guidelines. Basically, his approach has been to use as many areal equivalent profile parameters to the areal equivalent. Obviously, this helps to reduce confusion but some less obvious parameters he has suggested, called **functional parameters**, are based on common sense rather than proven usefulness. Some of these will be outlined below to illustrate the problems involved. A first important step has been to agree to designate areal parameters by the letter $S$ rather than $R$ for profile parameters, so that $S_a$ is the equivalent of $R_a$ and so one. The 'set' comprises:

1.   Amplitude parameters.
2.   Spatial parameters.
3.   Hybrid.                                     as in the profile
4.   Area and volume parameters.
5.   Functional parameters.

Table 3.5 gives the symbols for the 'conventional' parameters.

**Table 3.5**    *Conventional parameters – Eur 15178EN Provisional*

RMS of surface $S_q$
Ten point height $S_z$
Skew $S_{sk}$
Kurtosis $S_{ku}$
Density of summits $S_{ds}$
Texture aspect ratio $S_{tr}$
Texture direction $S_d$
Shortest correlation length $S_{al}$
RMS slope $S_{\Delta q}$
Mean summit curvature $S_{sc}$
Material surface area ratio $S_{ar}$

These are defined below.

### 3.4.3.1    Amplitude parameters

1.    Arithmetic average $S_a$ corresponding to $R_a$:

$$S_a = \frac{1}{L_1 L_2} \int_0^{L_1} \int_0^{L_2} \left| f(x,y) - \overline{f} \right| dxdy \tag{3.19}$$

where $\overline{f}$ is the height of the mean plane and $L_1$ and $L_2$ are the extent of the sample. $f(x,y)$ is the surface height at $x,y$. Also, $|\ |$ indicates that the sign is ignored.

2.    Root mean square $(R_q)$ value in 3D is $S_q$:

$$S_q = \sqrt{\frac{1}{L_1 L_2} \int_0^{L_1} \int_0^{L_2} \left( f(x,y) - \overline{f} \right)^2 dxdy} \tag{3.20}$$

3.    Skew $S_{sk}$:

$$S_{sk} = \frac{1}{L_1 L_2 S_q^3} \int_0^{L_1} \int_0^{L_2} \left( f(x,y) - \overline{f} \right)^3 dxdy \tag{3.21}$$

4.    Kurtosis $S_{ku}$:

$$S_{ku} = \frac{1}{L_1 L_2 S_q^4} \int_0^{L_1} \int_0^{L_2} \left( f(x,y) - \overline{f} \right)^4 dxdy \tag{3.22}$$

5.    Ten point height $S_z$:

$$S_z = \left( \sum_{L=1}^{5} \left| P_{i\,max} \right| + \sum_{L=1}^{5} \left| V_{i\,max} \right| \right) / 5 \tag{3.23}$$

### 3.4.3.2  Justification for functional parameters

Some advantages of using functional parameters are given below; they are subjective and so need to be validated. They have not been adopted as yet.

(a)  $S$ notation is usually used for stressed surfaces.
(b)  Functional parameters can be specifically matched to a given use.
(c)  Functional software is easier to understand?

Item (c) above is stated but not quantified.

Table 3.6 gives some of the functional parameters suggested. They are loosely tied to the material ratio curve, which makes sense because it links directly back to the profile.

**Table 3.6**  *Functional parameters*

| |
|---|
| Surface material ratio $S_q$ |
| Void volume ratio $S_{vr}$ |
| Surface bearing index $S_{bi}$ |
| Core fluid retention $S_{ci}$ |
| Valley fluid retention $S_{vi}$ |

**Table 3.7**  *Additional possibilities, some functional parameters*

| |
|---|
| $S_{bc} = S_q/Z_{0.015}$ |
| Where $Z_{0.015}$ is the height of the surface at 5% material ratio. |
| Core fluid retention index |
| $S_{ci} = (V_v(h = 0.8))/S_q(\text{unit area})$ where V is valley |
| If $S_{ci}$ is large then there is good fluid retention. |
| Relationship is $0<S_c<0.95 - (h_{0.05} - h_{0.8})$ |
| Valley fluid retention index |
| $S_{vi} (V_v(h = 0.8)) /S_q (\text{unit area})$ |
| $0<S_{vi} <0.2 - (h_{9.8}h_{0.05})$ |

**Table 3.8**  *Isotropy*

| |
|---|
| This can be found by: |
| *Longest bandwidth* = 1/isotropy ratio<br>Shortest bandwidth |
| *Shortest correlation length* = isotropy ratio<br>Longest correlation length |
| *Shortest average wavelength* = isotropy ratio<br>Longest average wavelength |
| *Shortest $S_m$* = isotropy ratio<br>Longest $S_m$ |
| The isotropy wave property of the surface has also been called the *texture aspect ratio*. |

Table 3.5 lists a few of the parameters and Table 3.6 gives some of the mathematical formulae for such parameters. They can be compared, where necessary, with the equivalent $R$ parameter derivations in Figure 3.3.

Table 3.6 gives the names of some of the functional parameters suggested and Table 3.7 some of the derivations.

Table 3.8 shows four possible ways of defining isotropy – lay or the lack of it. Many other attempts have been made to define the areal properties e.g. Peklenik and Kubo [3.6, 3.7]. The problems encountered in areal situations are many times more difficult than for the profile. Differences in definitions are possible, even for a simple parameter like isotropy. All of these can be related if the surfaces are random. It could be argued that any of these definitions would be sufficient in practice.

To be consistent, each parameter should itself be areal i.e. a plot of any of these isotropic possibilities.

Figures 3.43 and 3.44 give some of the graphical background to a conventional parameter (material ratio) and a suggested functional parameter (*void volume ratio* $S_{vr}(h)$).

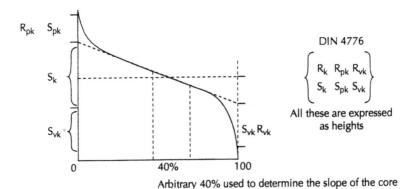

**Figure 3.43**   *Bearing ratio (material ratio)*

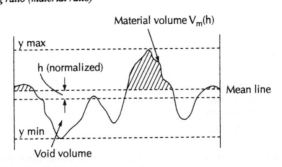

**Figure 3.44**   *Functional parameter*

Table 3.9 shows some of the parameters suggested up to now. This table shows a conservative set of parameters and even this is 17 in number.

Stout has, with some difficulty, reduced this to a basic set of 14 possibilities but even this number poses problems.

In Figure 3.43, the equivalent of the material ratio is considered and Figure 3.44 is an example of a function parameter.

$$
\left.
\begin{aligned}
Void\ volume\ ratio &= \frac{V_v(h)}{V_v(h_{max})} = S_{vr}(h) \\
Similarly\ V_m &= Volume\ (material) = S_{mr}(h)
\end{aligned}
\right\}
$$

Notice that these are not independent!

**Table 3.9** *Suggested parameters*

*One attempt to provide a limited set of areal (3D) parameters EUR 15178EN*

|  |  |  |  |  |  |
|---|---|---|---|---|---|
| (1) | $S_q$ rms of surface | | | | √ |
| (2) | Ten point height $S_z$ | amplitude parameters | √ | | |
| (3) | Skew $S_{sk}$ | | | | √ |
| (4) | Kurtosis $S_{ku}$ | | √ | | |
| (1) | Density of summits $S_{ds}$ | | | | √ |
| (2) | Texture aspect ratio $S_{tr}$ | | | | √ |
| (3) | Texture direction | spatial parameters | | | √ |
| (4) | Fastest decay of auto correlation $S_{al}$ | | | √ | |
| (1) rms slope $S_q$ | | | | √ | |
| (2) Mean summit curvature $S_{sc}$ | hybrid | | | √ | |
| (3) Developed surface area ratio $S_{dr}$ | | | | √ | |

+

*Some functional parameters*

e.g.
(1) Surface bearing area ratio $S_\phi$
(2) Void volume ratio $S_{vr}$
(3) Material volume ratio $S_{mr}$
(4) Surface bearing index $S_{bi}$    √
(5) Core fluid retention $S_{ci}$    √    **+**   valley fluid retention $S_{vi}$    √
(6) This is a set that mixes in some already existing parameters for the profile-extended to areal (3D) and some attempts to introduce the functional parameters.

There has been a proposal for the *14* ticked parameters to be taken as the primary set.

### 3.4.4 General comments on texture and its parameters

What has been set out above is the background to surface texture parameters for profile and areal representation of the surface. It is by no means exhaustive, yet it is already too large to fit into an easy standards scene. This suggests that somewhere along the way the basic thread has been lost.

In terms of manufacture, the picture is nothing like as bleak as it is in function. Basic parameters $R_a$, $R_z$ etc. can be used effectively in statistical process control, providing that not much meaning is attached to the actual value.

Process and machine monitoring can be effectively controlled by surface analysis using some of the random process analysis as has been shown. The only improvement likely in this subject area is the introduction of 'space-frequency' functions. These include Wigner distributions, ambiguity functions and wavelet theory. These can be used to detect not just the statistical information but changes in the statistics (Figure 3.45). These allow further investigation into machine tool performance and can even differentiate the mode of vibration of the tool on the workpiece. Whether or not this is cost-effective remains to be seen. A point mentioned earlier says that the surface of the workpiece is produced anyway. If there is any method of getting information from it, then it should be taken.

In addition to these 'space frequency' functions, there are new ways of characterizing surfaces based on 'fractal' theory. This will be discussed briefly in Section 3.5.

## 3.5 Space frequency functions

Conventional random process analysis cannot detect small changes in the statistics – for nano applications small changes can cause serious damage e.g. in wafer polishing. Extra machine tool monitoring can also be helpful. The following gives the formulae for the conventional random process function and the newer ones. These are listed here to show the similarities and differences. Compare:

Fourier spectrum 

$$F(w) = \int_{\infty}^{\infty} f(x)\exp(-jwx)dx$$

Autocorrelation function 

$$A(\beta) = \frac{1}{L-\beta} \int_{0}^{L-\beta} f(x) \cdot f(x + \beta)dx \qquad (3.24)$$

Wigner distribution 
$$W(x,e) = \int_{-\infty}^{\infty} f\left(x - \frac{\chi}{2}\right) \cdot f^*\left(x + \frac{\chi}{2}\right) \exp(-j\overline{w}\,\chi)dx$$

Ambiguity function 
$$A(\chi,\overline{w}) = \int_{-\infty}^{\infty} f\left(x - \frac{\chi}{2}\right) \cdot f^*\left(x + \frac{\chi}{2}\right) \exp(-jw\,\chi)dx$$

Both space frequency functions have the 'core' of the random process analysis but have one extra variable, which enables statistical change in the surface to be detected (Figure 3.46).

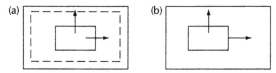

**Figure 3.45**    *Wigner and ambiguity functions (a) size change (b) position change*

## 3.6  Comments on digital areal analysis

There is another consideration to take into account when discussing the use of areal parameters in place of profile parameters. This is digital analysis. Some of the practical differences are shown below. Most important is the digital representation of summits (the areal equivalent of a peak).

Figure 3.47 shows various sampling patterns to describe a summit. Obviously, they are not the same as each other and they are not the same as defining a peak on the profile: there is much more flexibility, which can cause problems. The areal flexibility can at the same time be an aid to understanding or means of generating confusion.

### Wavelet filters (space frequency)

The fundamental idea is to analyse the waveform according to *scale* (or resolution).

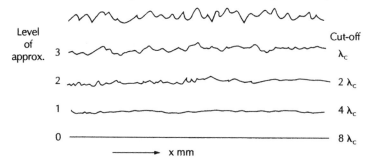

**Figure 3.46**    *Wavelet – mathematical zoom lens*

It is equivalent to an octave band filter bank, each one using the same shape 'wavelet' – but each of different scale.

Originally, the wavelet was a Gaussian pulse but now all shapes can be used. The difficulty is to make the system orthogonal i.e. each band-independent.

There is potential for using the wavelet method with fractals, as both use different scales.

The actual identification of a summit depends on how it is defined. There are a number of ways the summit can appear in digital form (Figure 3.47).

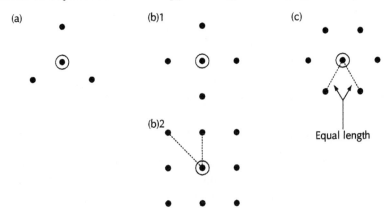

**Figure 3.47** *Digital forms of summit appearance*

| (a) | Triagonal | | 3 points at 120° must be lower than the centre one |
|-----|-----------|------|---------------------------------------------------|
| (b) | Rectangular | b(1) | 4 points surrounding centre one |
|     |           | b(2) | 8 points surrounding centre one |
| (c) | Hexagonal | | 6 points surrounding centre one |

**Note:** b(2) has unequal spacings relative to the centre one i.e. the correlations are unequal.

**Note:** the parameter values do not converge to the profile values as the spacing is reduced.

Hence, parameter values depend on the numerical model as well as on the spacing [3.10]. Figures 3.48, 3.49 and 3.50 show some other problems.

Areal sampling is not straightforward. Take, for example, triagonal sampling. In

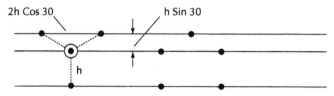

**Figure 3.48** *Triagonal sampling*

this, the sampling needs to have different spacings between tracks and different start times per track, which is instrumentally possible but can be messy.

## Numerical convergence

Take the case of a profile peak:

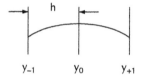

**Figure 3.49** *Convergence of peak definition*

There is no ambiguity. Now, as h reduces to zero the value of, say, the curvature converges to the continuous or theoretical value.

## Case of areal summit

The rectangular sampling here is an example.

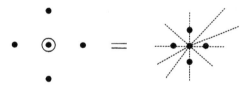

**Figure 3.50** *Non-convergence of summit definition*

It can be seen that it is possible to miss a peak that is at a diagonal to the grid pattern. How much information is missed depends on the numerical model. Using more points to make up a more complicated Lagrangian model reduces the risk of missing a peak but there is always a finite possibility of peaks (summits) being missed [3.9].

## 3.7 Two-dimensional filtering (areal filtering)

Two-dimensional filtering has two spatial dimensions instead of one, yet it is sometimes wrongly called 3D filtering. As with digital sampling, the 2D filtering has cross terms that are not seen easily from profile filtering. The filtering action is that of convolution, which has already been mentioned in the space domain corresponding to multiplication in the frequency domain. In this case, the window or weighting function itself has two spatial dimensions $x$ and $y$ as indicated in Figure 3.51.

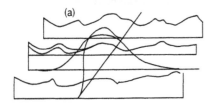
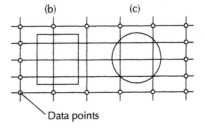

Data points

**Figure 3.51**   *Areal sampling*

To get best results, the weighting function boundary shape should be as close as possible to the lay pattern. This is a constraint but it is only weak because the convolution can take place anyway. Ideally, for manufacturing the weighting function, it should have some connection with the path of the tool.

The basic equation for the 2D spectrum is $F(w,v)$

$$F(w, v) = \int\limits_{-\infty}^{\infty} \int\limits_{-\infty}^{\infty} f(x, y) \exp(-(wx + vy)) dx dy \qquad (3.25)$$

and the autocorrelation

$$A(v, \gamma) = \int \int f(x, y) f(x - \tau, y - \tau) dx dy \qquad (3.26)$$

The main thing to remember is that there are two spatial axes x and y. This is not to be confused with the two variables in the Wigner distribution. In these, one of the variables is spatial $x$, one is frequency $w$ i.e. only one spatial variable. The problem is that the 'space frequency' functions are plotted against two axes in exactly the same way as the areal (3D) autocorrelation but the information is still about the profile and not the areal surface.

In summary, two-dimensional filtering gives more information about the spatial character of the surface. Space frequency functions give a more comprehensive picture of the changes in the nature of the surface.

Note about space frequency functions – the functions above have two variables $x$ and $w$. These are used for analysing the surface geometry. However, they can and were originally used as 'time –frequency' functions. In this mode, $x$ is replaced by $t$ and $w$ is the temporal version of $w$. In this mode, real time machine vibration can be examined. There is no reason why both versions could be used: the spatial form for surface characterization and the temporal for real time vibration problems in the machine tool.

## 3.8 Fractal surfaces [3.11]

From what has been said, it is clear that the autocorrelation and power spectrum can be used in surface analysis. One fundamental type of spectrum is given below for, say, a ground surface.

$$P(w) = \frac{K}{w^2 + a^2}$$

Fractal surfaces have a similar spectrum

$$P(w) = \frac{K}{w^v}$$

$v$, **the index for fractal surfaces**, can take values between 0 and 2 allowing non-integer values. From this index, one of the two fractal parameters can be obtained. Thus $+ v = 5-2D$; the other parameter called the **topothesy** $\Lambda$ can be obtained from the structure function [3.11].

There is a big debate at present about the relevance of fractals in surface metrology. Why bother? The essential property of a fractal surface is that the parameters D and $\Lambda$ work out to be the same whatever the scale – a bit like 'fleas on fleas' as seen in Figure 3.52. All conventional parameters like curvature, slopes etc. are very scale-sensitive. Using conventional geometry, there is no such thing as a number or numbers representing the surface.

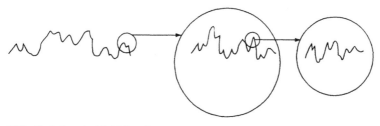

**Figure 3.52**   *Scale invariant fractal surface*

So, the idea is that with fractal geometry under certain conditions, a unique number can be assigned to the surface. At first sight, this appears to be a breakthrough in surface characterization but there is a snag. This is that functional uses of surfaces are very scale-dependent as will be seen later. Consequently, surface parameters hoping to predict performance should be equally scale-dependent. In other words, fractal geometry is precisely what is not required. Fractals tend to mask sensitivity to the manufacturing parameters and functional situations. It seems that fractal geometry, although elegant, is not relevant in manufacture [3.11].

## 3.9 Summary of characterization

Figures 3.53 and 3.54 show what can be used as the basic elements that should be used in characterization. Notice that it is not possible to isolate the characterization of the surface from either the manufacture or the workpiece function. It is this link that has been missing. Without these two constraints, surface characterization can get out of hand.

The large number (14) of even the simplest set of areal parameters make it unacceptable in use. It is meaningless to advance further down the road of discrimination without pausing to consider interactions with other parts of the system. This approach should be more fruitful.

Some elements of this interdependent approach are given below. The fact that each station can conveniently be split into two rather obvious factors is a tremendous bonus.

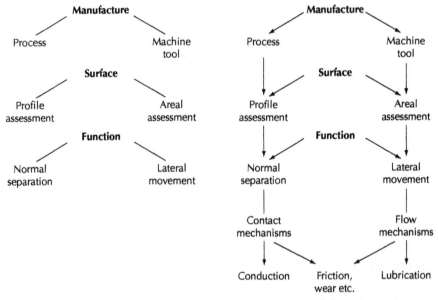

**Figure 3.53**  *Basic characterization*          **Figure 3.54**   *Elementary links in characterization*

At the functional level, it is possible to expand the characterization into conduction and lubrication as well as contact and flow. Hence a $2^4$ classification, which is just about the *maximum* the designer can work with and the *minimum* the tribologist will allow.

## References

3.1  Abbott E. J. and Firestone F. A. *Specifying Surface Quality*. Mech. Eng. Vol. 55 p569–572.

3.2  Pesante M. *Determination of Surface Roughness Typology by Means of Amplitude Density Curves*. Ann. CIRP Vol. 12 p61 (1963).

3.3  Ehrenreich M. *The Slope of the Bearing Area as a Measure of Surface Texture*. Micro Technic XII p83 (1959).

3.4  Whitehouse D. J. *The Parameter Rash Is There a Cure?* Wear 83 p75–78 (1982).

3.5  Stout K. G. *Development of a Basis for 3D Surface Roughness*. EC (1996) Contract No. SMT4-CT98-2256.

3.6  Peklenik J. and Kubo M. *A Basic Study of the Three Dimensional Assessment of the Surface Generated as a Manufacturing Surface*. Ann. CIRP Vol. 16 p257 (1968).

3.7  Kubo M. and Peklenik J. *An Analysis of Micro Geometric Isotropy for Random Surface Structures*. Ann. CIRP Vol. 16 p235 (1968).

3.8  Raja J. *Filtering of Surface Profiles – Past, Present and Future*. Precision Engineering Proc. 1ˢᵗ Int. Conf. on Prec. Eng. p99 (2000).

3.9  Najak P. R. *Random Process Model of Rough Surfaces*. Trans. ASME Vol. 93 p398 (1971).

3.10 Whitehouse D. J. and Phillips M. J. *Two Dimensional (Areal) Properties of Random Surfaces*. Philos. Trans. Roy. Soc. A 305 p441–448.

3.11 Whitehouse D. J. *Fractal or Fiction*. Wear 249 p345–353 (2001).

# 4

# Surface metrology and manufacture

## 4.1  Where and when to measure

Using the surface texture as a check on the process is well established but there are a number of issues to consider. These are: where to measure, what to measure and when to measure.

Ideally, the measurement should take place as the workpiece is being made, so as to avoid scrapping more than one part. This in-process possibility is very difficult to achieve because of the hostile environment in which the part is made. Often. it is the presence of metal chips and/or coolant spray that makes measurement difficult but there is a trend towards dry cutting, which may help alleviate this problem. The number of parameters that could be measured and displayed in-process is limited although robust surface-measuring equipment cannot be expected to perform well.

Another possibility is to have the surface tested when the part has been made but not moved. The surface instrument, which is hand-held, has somehow to be perched on the part when the machining has stopped and then the measurement recorded.

Alternatively, the part could be removed from the machine (the lathe, say) and measured with an instrument located near to the machine tool. This is usually called in situ measurement and the availability of instrumentation with more parameters is much higher than in the in-process case.

It would normally be the case that the in situ tests are carried out during every shift. The tests in the cases where the measurement is on or near the machine will be concerned with the process.

For more rigorous tests involving perhaps roundness as well as roughness and waviness, the workpiece would be taken to a properly equipped inspection room. Here, some tests for machine tool capability would be possible.

Failing this facility, the tests could be carried out at a 'centre of excellence' in which personnel as well as equipment are 'capable'.

## 4.2   The process and surface finish

### 4.2.1   General

Table 4.1 shows typical surface roughness values for commonly used machining processes. These processes can be categorised into a number of subdivisions.

**Table 4.1**   *Typical roughness values produced by processes*

| Process | Roughness values (µm $R_a$) |
| --- | --- |
| | 50 · 25 · 12.5 · 6.3 · 3.2 · 1.6 · 0.8 · 0.4 · 0.2 · 0.1 · 0.05 · 0.025 · 0.0125 |
| Flame cutting | |
| Snagging | |
| Sawing | |
| Planing, shaping | |
| Drilling | |
| Chemical milling | |
| Electro-discharge machining | |
| Milling | |
| Broaching | |
| Reaming | |
| Boring, turning | |
| Barrel finishing | |
| Electrolytic grinding | |
| Roller burnishing | |
| Grinding | |
| Honing | |
| Polishing | |
| Lapping | |
| Superfinishing | |
| Sand casting | |
| Hot rolling | |
| Forging | |
| Permanent mould casting | |
| Investment casting | |
| Extruding | |
| Cold rolling, drawing | |
| Die casting | |

Key: ■ average application   ▨ less frequent application

1.   Cutting with single or multiple tool tips – this includes turning, milling, broaching, planing.
2.   Abrasive machining – this includes grinding, polishing, honing.
3.   Physical and chemical machining – this includes electrochemical machining, electrodischarge machining etc.
4.   Forming, casting, extrusion.
5.   Other macroscopic machining to include laser machining, high-power water jet.
6.   Ultra-fine machining (nanomachining) including ion beam milling and energy beam machining.

Within these categories, there is considerable overlap, for example, diamond turning is capable of producing very fine surfaces, yet it is included here in group 1. In fact, the groups have been assembled according to generic likeness rather than magnitude of roughness, because it is the former that impart the unique character of the process to the workpiece.

### 4.2.2   Turning

This is the commonest process and, as the name implies, involves an axis of rotation. Some turning modes are shown in Figure 4.1 [4.1].

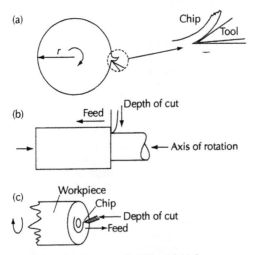

**Figure 4.1**   *Modes of turning (a) radial (b) axial (c) face*

Typical variables are cutting speed – workpiece peripheral speed relative to the tool axial feed – the advancement of the tool per revolution of the workpiece, the shape of the tool and the depth of cut of the tool into the workpiece material.

There are other very important aspects that are not shown on the diagrams but which contribute a considerable difference to the form and roughness. These include the absence or presence of coolant and, if present, its constitution and the method of supply, whether fluid, mist or drip, and so on. In addition to these effects is the effect of the machine tool itself. But in general, the actual value of the roughness can be estimated at least crudely in terms of height and form from a knowledge of the basic process parameters, unlike the case for grinding.

As far as turning is concerned, it is rarely (except in the case of diamond turning) used for very fine finishes. It is a very useful method for removing stock (material) in order to produce the basic size and shape. As a general rule, the surface roughness tends to be too rough to be used in very critical applications in which high stresses could be detrimental, but nevertheless there are many applications where turning (single-point machining) is used because of the speed and efficiency of the process.

## *Theoretical surface finish – effect of tool geometry*

Figure 4.2 shows a typical cutting tool with the primary and secondary cutting edges.

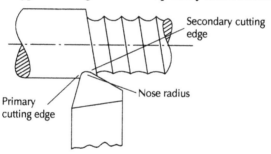

**Figure 4.2**   *The two cutting edges*

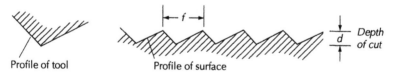

**Figure 4.3**   *Surface produced by angular tool*

In its simplest form, the tool can be considered to be a triangle (Figure 4.3). If $f$ is the feed and $d$ is the depth of cut, then:

$$R_t = d \text{ and } R_a = \frac{d}{4}$$

The roughness value is independent of the feed $f$. The triangular shape is impractical; in most cases the tip is curved, having a nominal radius $R$. The typical profile is shown in Figure 4.4. This assumes that the cutting is confined to the curve of the tool.

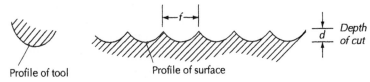

Profile of tool          Profile of surface

**Figure 4.4**   *Surface produced by curved tool*

The roughness can be given by:

$$R_T = r - \sqrt{r^2 - f^2 / 4} = r\left(1 - \sqrt{1 - f^2 / 4r^2}\right) \tag{4.1}$$

$$R_t \sim f^2 / 8r \tag{4.2}$$

The feed is very important in this case. $R_t$ is proportional to $f^2$. $R_a$ is approximately given by:

$$R_a \sim 0.03 \frac{f^2}{R} \tag{4.3}$$

Again, the $R_a$ is about a quarter of the peak to valley $R_t$. In fact, this ratio $R_t/R_a$ can be taken as 4:1 for single-point cutting.

These simple formulae for $R_t$ and $R_a$ give 'theoretical' surface finish in terms of the process parameters. In practice, the finish is less because of other factors such as tool wear, built-up edge and fracture roughness.

It is interesting to recall that if the profile is a true sine wave, the ratio is $\pi$:1 and if it is a square wave it is 1:1; neither of which occur in normal turning.

The roughness values are usually taken to be in the axial direction produced by the secondary cutting edge as shown in Figure 4.2. Primary cutting edge roughness is usually in the circumferential direction, in which case the finish can be masked by out-of-roundness errors caused by chatter or other machine tool problems.

The roughness produced in both cases is very dependent on the cutting speed i.e. the surface speed relative to the tool. Secondary edge roughness quickly approaches the theoretical value if the speed is increased.

Fracture roughness [4.2] degrades the surface. It occurs at slow speeds and affects the subsurface with the result that chip fracture occurs. Above 1m/sec, this effect disappears. **Built-up edge (BUE)** is caused because of friction between the chip and the tool. At some point, the friction will be large enough to cause a sheer fracture in the chip in the region of the tool face. This causes some deposition on the tool face, which in turn increases the friction and so on until the chips begin to weld to the tool, thereby degrading the cutting action and the resulting surface finish. There are various ways of getting rid of the BUE problem. One is to introduce more coolant and/or increase the cutting speed.

Figure 4.5 shows the boundary of BUE with cutting speed and tool-chip interface temperature.

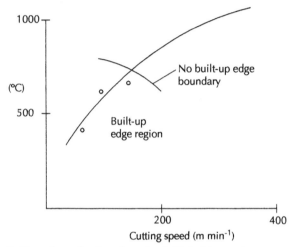

**Figure 4.5**   *Built-up edge as function of speed and chip tool temperature*

Ideally, the tool tip shape replicates itself in the surface as will be seen. In fact, it is the negative of the tool tip, so in principle the surface finish can be used to measure tool wear. There is some uncertainty of the geometry of the cut at the trailing edge. Also, the metal at the trailing edge of the tool is subject to high stresses and tries to flow to the side in order to relieve this stress.

Other effects of machining on the surface geometry include what is called **spanzipfels** [4.3]. This effect is due to the minimum undeformed ship thickness, which allows a small amount of material to be left on the surface instead of being removed by the tool. The theoretical surface finish is slightly worsened. The $R_t$ becomes:

$$R_t = \frac{f^2}{8R} + \frac{t_m}{2}\left(1 + \frac{Rt_m}{2}\right) \tag{4.4}$$

where the second term is the spanzipfel. It is usually small but can be up to 20% of the roughness in special cases such as dry turning. This is worrying in view of the increasing trend towards dry working. A typical value of $t_m$ is 1μm.

There are many other factors that influence the surface finish. For example, the initial sharpness of the tool influences wear, which in turn degrades the roughness. Cutting fluid is important, especially at low speeds, where the friction can be high. This is not a marginal effect. Factors of 4:1 in roughness improvement issuing effective cutting fluids have been reported.

Inhomogeneity of the material in the form of inclusions and voids disturb chip formation and flow, hence the surface. It should also be noted that the process of turning should direct the chips away from the workpiece so as not to mark the surface.

Tool wear has been mentioned earlier in terms of the cutting edge. Actually, there are three regions on the tool that can wear. These are shown in Figure 4.6. Here, region A is the tip itself, region B is the tool flank and region C is the tool top. In regions A and B, scratches and fracture are most likely to occur. In region C, the principal type of wear is due to cratering, which in turn affects the possibility of the formation of BUE. Each of these regions affects different regions of the profile as seen by an instrument.

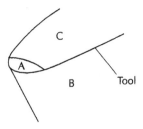

**Figure 4.6** *Wear regions of tool*

### 4.2.3 *Diamond turning*

This is quite different from ordinary turning in the sense that a complicated tool shape is often involved as well as there being a complicated cutting mechanism [4.4]. Diamond turning is frequently used for obtaining very fine surfaces on soft materials such as aluminium and copper, where abrasive methods like grinding are unsuitable. Furthermore, diamond turning is carried out at very high cutting speeds so that problems associated with BUE do not arise. Under these circumstances, it should be possible to achieve the true theoretical surface roughness. On the other hand, the cutting tool can have two facets, one that cuts and one that can be used to burnish the machined surface (Figure 4.7). When the burnishing follows the cutting, the surface finish can be very fine indeed. Typical roughnesses for ordinary diamond turning may be about 25-50nm $R_a$, but with burnishing they can be considerably reduced.

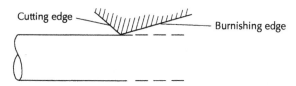

Cutting edge

Burnishing edge

**Figure 4.7** *Dual purpose of cutting tool*

However, the smoothing over of the tool mark by plastic flow caused by the following edge can introduce detrimental physical properties into the surface such as residual stress, so the burnishing can only be used in non-critical stress applications or where relaxation caused by the presence of stress could deform the workpiece.

The amount of burnishing also has an upper limit. If it is larger than necessary, then the height levels of the crystal grains will differ. This is due to the different crystallographic orientations of the grains.

### 4.2.4 Milling and broaching

At first glance, milling and broaching appear to be quite different processes. This is not altogether so because surface broaching is similar to peripheral milling, having a cutter radius of infinity. In fact, surface broaching is being increasingly used to replace milling as a stock removal process and at the same time milling is being used for finer surfaces.

At one time, milling was considered to be another form of sawing: in effect a two-dimensional saw. It has a fundamental difference from turning because the cutting is intermittent. The big problem is ensuring that all the teeth on the milling cutter actually remove material, otherwise those that drag have a tendency to work-harden the surface, which then destroys the edge of the working teeth. Also, if not enough mechanical support is given to each cutting edge, the tip can be torn or moved from its seating. The impulsive form of machining in milling makes it much more complicated than other forms to analyse.

Milling processes are many and varied and, as a result, the cutters used and their impression on the surface can be complex. However, there are two general types: **peripheral milling** and **face milling**. Other applications such as the milling of slots, forms and helixes are essentially variants of the basic methods and each is specialized to such an extent that general surface considerations are not possible.

In peripheral milling, the cutter teeth are machined to give cutting edges on the periphery. These may be gashed axially or, more often, spirally as seen in Figure 4.8. There are two rake angles, one for the spiral and the other for the axial raked. It can easily be shown that a large effective rake can be produced on a milling cutter without unduly weakening the tooth by having a large radial rake. This is important because, as in turning, the cutting force per unit length of cutting edge reduces rapidly with an increase in rake. Another advantage of using a helical rather than a purely radial cutter is a more even distribution of cutting force, which tends to preclude chatter and hence waviness on the surface. A disadvantage is the end thrust component of the cutting force.

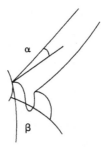

**Figure 4.8**   *Angles of importance in milling: α, spiral angle; β, axial rake*

There are two methods of removing metal, one known as **upcut milling** and the other as **downcut** or **climb milling**. The essential features are shown in Figure 4.9.

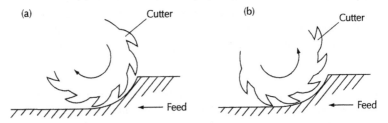

**Figure 4.9**   *Milling cuts (a) climb milling (b) upcut milling*

Although upcut milling is most often used, it does rely on very efficient clamping of the part on the bed, otherwise chatter is easily generated on the surface. Climb milling does not need this but does require an adequate backlash eliminator [4.5]. Another difference is that the two techniques produce different shaped chips.

Typical pictorial representations of upcut and downcut milling are shown in Figure 4.9. Peripheral milled surfaces in general are not critical or are finished with another process.

The other main type of milling – face milling – is capable of producing finer surfaces (Figure 4.10), so more detail as an example will be given here [4.6]. The factors affecting the surface texture in face milling are as follows.

1.   Selection of cutter geometry.

2.   The accuracy of grinding of the selected angles on the cutter teeth.
3.   The setting of the teeth relative to the cutter body.
4.   The alignment of the machine spindle to the worktable.

The effect of imperfect milling reveals itself on the surface dramatically. This is apparent in the lay pattern, which is very visible.

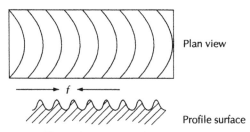

**Figure 4.10**   *Surface appearance of face milling single tooth*

To consider these points in the way described by Dickenson [4.6], a selection of cutter angles is shown in Figure 4.11. The various tool angles on a face-killing cutter fall into two categories: (i) the cutter geometry, which directly affects the surface roughness and (ii) those angles that indirectly affect the surface roughness by influencing the cutting actions. The radial clearance angle and the corner angle fall into the first of these two categories.

It seems from tests that the surface roughness values deteriorate as the radial clearance angle is increased, so to keep a good surface finish this should be kept small. If a corner tip radius is used, it should be approximately 10μm.

There are a number of key issues that determine the roughness.

(a)  *Setting of teeth relative to cutter body*

Anyone who has examined face-milled parts will realize that the dominant mark on the surface is invariably produced by one or two teeth being non-coplanar with the rest. This has the effect of producing a periodicity corresponding to feed per revolution superimposed on the feed/tooth periodicity. In theory, the surface roughness should be determined from the latter; in practice it is more likely to be the former.

The fact is that if one tooth is only a very small amount lower than the other teeth, it completely swamps out the other cutter marks. As an example [4.6], on a 12-tooth cutter with a radial clearance angle of ¼° operating at a feed of 0.1mm/tooth, one tooth would need only to be proud by 0.5μm! This illustrates the sensitivity of the setting. Also, the fact that the dominant periodicity is changed from feed/tooth to feed/revolution means that the cut-off wavelength of the instrument has to be chosen with care. In the example above, a cut-off of 2.5mm would

(a)

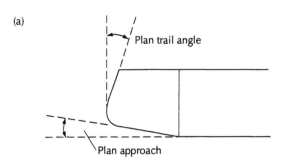

(b)

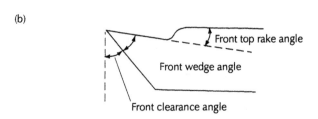

(c)

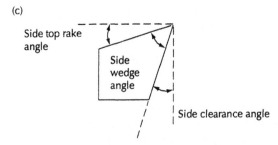

**Figure 4.11**   *Views of tool with relevant angles: (a) plan view (b) front (c) side view*

be more than adequate but 0.8mm cut-off would give disastrously low readings of roughness. This is because the dominant periodicity would lie outside the cut-off (i.e. 1.2mm), which means that the true roughness value is attenuated.

In practice, there would be no problem because the inspector should examine the surface prior to measuring and determine the suitable cut-off relative to the dominant periodicity as laid out in the standards.

### (b) *The pattern of roughness on the surface*

This is determined to a great extent by the alignment of the spindle to its worktable. Any misalignment will lead to an increase in the surface roughness values for cutters that are ground without a corner radius. Obviously, if the cutters have a radius, it is not so important.

The pattern, assuming that no one tooth is proud, can take three forms.

1.   Where all the leading edges of the cutters generate the mark – all called **front-cut**.

2.   Where all the back edges of the cutters generate the mark – all called **back-cut**.

3.   Where there is equal cutting front and back.

Ideally, the third possibility is the one that should be produced but, in practice, it is difficult to avoid back cutting because of the run-out of the spindle and the elastic recovery of the material.

As with the 'proud' tooth problem, the main issue in measuring roughness in face-milling alignment problems arises with the selection of the cut-off. Unbelievably, the problem is worst when the front cutting and back cutting are more or less balanced, as shown in Figure 4.12.

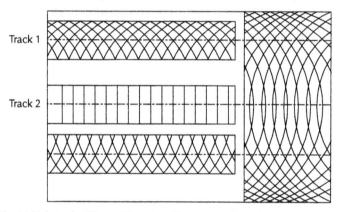

**Figure 4.12**   *Multiple tooth milling – different profiles, same surface*

The spatial interaction of forward cutting and backward cutting can produce some very strange periodic effects if only one profile is taken. Shown in Figure 4.12 are two tracks. The waveform of the periodicity is quite different in both cases, as is the variation in the spacing of the dominant features. This means that it is quite difficult to decide which cut-off to use. Basically, the only method is to take a profile graph and ensure that at least five or so maxima lie within one sampling length. The difficulty arises because in such a roughness, one profile graph cannot adequately represent the surface. The variation of $R_a$ with position across the surface is quite staggering when it is considered that essentially the surface is very deterministic and well behaved.

It is not unusual for 40% variation to be found in the $R_a$ value on a regular face-milled surface owing to the positioning of the profile track relative to the centre position of the cutter path. Thus, at least three measurements on face-milled surfaces have to be taken. Note that if $R_q$ rather than $R_a$ or peak measurements were measured, the variation would be much less because the phase effects of the different components that cause the variations would not contribute.

### (c) Effect of feed on surface roughness

An increase in feed tends to increase surface roughness, so that feed should be as small as possible. However, the purpose of the milling operation is usually to remove metal as quickly as possible and so feeds tend to be high. As usual, the two requirements go against each other.

### (d) Other factors in milling

Lobing in radial peripheral milling caused by chatter is often due to the fact that all teeth on the cutter have the same directional orientation, which can in certain circumstances set up chatter. It has been shown, however, that some improvement in stability can be achieved if the cutters have non-uniform pitch, as illustrated in Figure 4.13 [4.7].

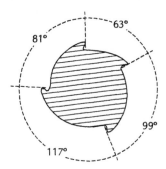

**Figure 4.13**   *Chatter-suppressing cutter*

### (e) Other types of milling

Ion milling is something of a misnomer and will be dealt with in superfine finishing.

Because milling is not normally regarded as a finishing process, relatively little work has been carried out on attempting to predict the surface finish from the process parameters – unlike in turning, only very early work has been reported [4.8].

The $R_t$ value of the surface roughness has been given by:

$$R_t = \frac{f^2}{8[R \pm (fn / \pi)]} \tag{4.5}$$

where $f$ is the feed per tooth, $R$ is the cutter radius and $n$ is the number of teeth on the cutter. The plus sign in the denominator corresponds to upmilling while the negative sign corresponds to downmilling. Equation (4.5) can be in good agreement, with practice, if the spindle of the milling machine is held to a very low value of run-out. It represents only a very simple modification of the theoretical surface roughness for turning. It seems probable that much more work needs to be done on this in the future with the renewed interest in milling.

Boring is a cutting operation intended for internal work and is used to produce the required dimension and shape to bores. The operation is, in some ways, closer to turning than milling because the cutting action is continuous. However, the feed is longitudinal in the axial direction. Also, differently shaped cutting tools may have to be used to ensure that the cutting edge is in constant pressure with the bore.

Broaching is a method of altering the size or shape of internal work by pushing or pulling a tapered cutting tool (broach) through the workpiece. The teeth are progressively increased in size towards one end of the broach. Many shapes of internal work or holes can be produced by this method. Keyways are commonly formed by broaching. The surface marks are axial and are difficult to measure if the hole is non-circular. In common with other cutting tools, grinding of the cutting edges is of prime importance because not only do the cutters have to be ground to the correct angle, but the pitch between the teeth as well as the taper of the tool has to be accurately preserved, otherwise unequal cutting forces occur.

Reaming and drilling are other processes using multiple teeth but the surface finish is usually not a critical factor, although the dimension is.

### 4.2.5   General abrasive process and surface finish

Unlike the previous processes involving single and multiple teeth in which the relationships between cutting edges is deterministic, abrasive processes involve a random distribution of grains whose edges act as cutters.

In the previous section, single-point and multiple-point cutting have been considered from the point of view of surface generation. In general, these processes are used for the removal of bulk material to get the workpiece down to its intended dimension. Single-point cutting can be used as a finishing process, particularly if the tool is a diamond. The advantage of having one tool is that it can be positioned more or less anywhere in space and so can produce many shapes. An alternative way to produce a finish is by means of abrasive processes. The basic difference is that there are multiple random grains that carry out the machining process rather than one fixed one (or a few in the case of multiple-tooth milling). Consequently, it is more meaningful to consider *average* grain behaviour rather than the individual behaviour of the grains. Put in terms of turning, it means that there are many tools cutting with negative rake at a very small feed. (This is equivalent to the effective grain width of the abrasive!) This ensures a much finer finish in itself. When coupled with the averaging effect of the whole wheel profile, the resultant roughness is very low. Also, whereas in turning and milling the cutting is continuous, in most abrasive processes it is not. The abrasive action is usually analysed in terms of three mechanisms – cutting, rubbing and ploughing – and, more importantly, the angle the grain makes with the surface is negative rather than positive as in turning. This negative rake angle makes the probability of burnishing and ploughing more prevalent than in straightforward cutting. There are many factors in the abrasive process that influence roughness and are not considered here, such as hardness etc.

There are three main abrasive processes: grinding, honing and lapping. The main difference between these is that in honing, although the grains are held rigidly, the speeds are low and reciprocating movement is often used. Superfinishing is similar except that a slight axial vibration is superimposed. Lapping is again slightly different in that the grains are loose, they are contained in a fluid and are therefore free to move. The type of surface produced by these methods is easy to identify as shown in Figures 4.14 and 4.15. It is basically in the areal view that the differences become obvious.

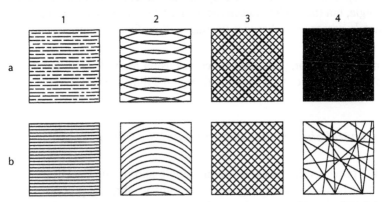

**Figure 4.14**  *Surface roughness (from DIN 4761): 1, peripheral grinding; 2, face grinding; 3, honing; 4, lapping; (a) and (b), process variations*

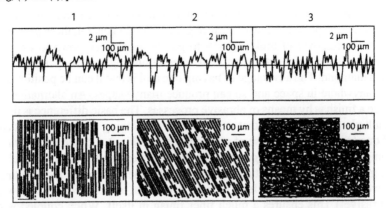

**Figure 4.15**  *Abrasive processes, same R, roughness of 4μm: 1, grinding 2, honing; 3, lapping*

Figure 4.16 shows the conventional grain sizes and the surface roughness produced by the true abrasive processes [4.9].

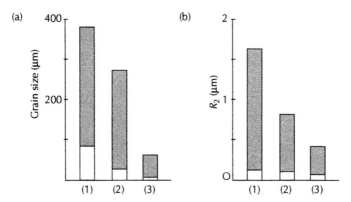

**Figure 4.16**    *Graph showing range of grain size and workpiece roughness: (1), grinding; (2), honing; (3), lapping*

The grain size in lapping is smaller than in the other processes as seen from Figure 4.16. It seems that the relationship between grain size and roughness is about 200:1 and the relation between $R_z$ and $R_a$ is about 6:1.

The fixed grain methods have repetitive process times but lapping does not. This implies that there is a further random element that should be taken into consideration for lapping. Although the contact versus process times are not the main factor influencing the surface geometry, the times greatly influence temperatures and subsurface effects.

## 4.2.6    Grinding and surface finish

Grinding wheels are made up of a large number of grains held by a bonding agent. The abrasive grain can be aluminium oxide, silicon carbide, cubic boron nitride or diamond (either natural or synthetic). The relative merits of these materials or the bonds will not be of primary concern in this context, except inasmuch as they affect the form or texture. The features of the wheel structure that are important in determining the resultant surface roughness are the density of grains/mm$^2$, the height distribution of the grains, the elasticity of the grains within the bond and the fracture characteristics of the bond and grain. Other associated issues include wheel dressing and spark-out. Because each of these has to be taken with the other parameters of importance such as cutting speed, workpiece speed and feed rate, it is more meaningful to discuss roughness and form in a number of specific configurations such as those shown in Figure 4.17.

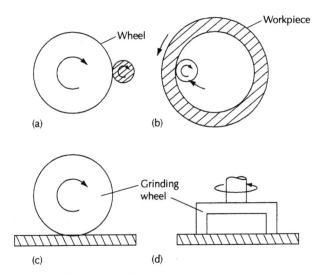

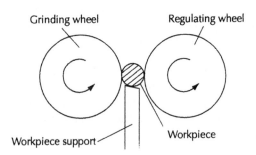

**Figure 4.17**   *Configurations of roughness and form*

**Figure 4.18**   *Centreless grinding*

The surface finish i.e. the $R_a$ for these variants is usually between 0.1 and 0.3μm and the peak to valley between four and six times the $R_a$. Note that the ratio of $R_t$ / $R_a$ is higher for abrasive processes than for single- and multiple-tooth machining.

Another difference is that extreme peaks in the abrasive processes are sharper than these in single-point cutting. So, in tribological situations, the behaviour can be significantly different. For example, in the case where two bodies touch as in a mechanical seal, the initial contacts of ground surfaces are likely to be plastic, whereas for cutting processes they are more likely to be elastic.

Centreless grinding is an important form of grinding used in making rollers for bearings etc. A layout is shown in Figure 4.18. Notice that the reaction by the grinding wheel on the workpiece has to be taken up by a regulating wheel that is also rotating. This complicates the process. Figure 4.19 shows how the various components of centreless grinding affect the finish of the work.

In order to isolate the features of this very important type of grinding that are most likely to influence roughness and roundness, it is useful to be reminded of the possible factors. The factors can be roughly divided into three categories, A, B and C, as in Figure 4.19 [4.10].

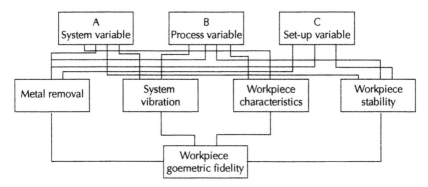

**Figure 4.19**    *Transfer of variables to workpiece geometrical fidelity*

**A.** The system variables (which can be assumed to be inherent) are:
1.   Rigidity of wheel and regulating wheel spindles and housings and the rigidity of the work rest holder.
2.   Specification of the grinding wheel (which determines the contact stiffness and removal rate).
3.   Grinding wheel speed – usually fixed.
4.   Vibration of two wheels and blade.
5.   Diameters of two wheels.
6.   Workpiece characteristics, hardness, geometry, length, diameter etc.

**B.** Process variables (easily changeable during grinding):
1.   The regulating wheel speed (controls the work speed, the through feed rate and the amplitude and frequency of regulating wheel vibration).
2.   Metal removal rate – infeed per workpiece revolution.
3.   Dressing conditions for grinding and regulating wheels.

**C.** Set-up variables – changed every set-up for a given workpiece:
1.   Work support blade – material, thickness and top-face angle.
2.   Work centre height above grinding wheel centre.
3.   Regulating wheel through feed angle.
4.   Diamond dresser angle with respect to the regulating wheel axis.
5.   Dressing diamond offset.

6.   Longitudinal profile of the regulating wheel dressing bar.

In view of the number of variables, they have been grouped under the following headings.

(a) Metal removal process.
(b) System vibration.
(c) Workpiece characteristics.
(d) Workpiece stability.

The transfer from the machine tool to the workpiece is thereby simplified.

Among the variables listed above, there are some relating to the set-up of the grinder; in particular, they are concerned with the regulating wheel and the work support blade. These two machine elements have no parallel in other examples of grinding and it is not surprising, therefore, that they are particularly important where roughness and roundness are concerned. Roundness especially is the feature of surface geometry that is most affected in centreless grinding.

It is most important that the regulating wheel is accurately trued. Conventionally, this is achieved by a single-point diamond dresser applied two or three times at a depth of 20μm and then at a light cut. Run-out of the regulating wheel periphery at the truing point is typically about 8μm, although that of the spindle itself is considerably better, as low as 0.2μm. Alternatively, the regulating wheel is ground by plunge grinding. This reduces the run-out and profile out-of-straightness to one-tenth of the conventional method. It is assumed that the slide accuracy of the truing grinder has the same order of accuracy as the machine slide or better.

Truing the regulating wheel in this way reduces its wear and reduces the surface roughness of the workpiece – say, cylinder rollers from 0.3μm $R_a$ to 0.1μm $R_a$. Even more dramatic is the improvement in roundness. A typical value of roundness error of 2μm can be reduced to 0.2μm when the regulating wheel is properly trued. All this is achieved at the cost of having a much stiffer system and a much narrower stable region [4.10].

Simulations have shown that, irrespective of the work support angle, when the workpiece centre height above the grinding wheel centre is zero, the system improvement as far as roundness is concerned is marginal, and that for workpieces with roundness values to be less than 0.5μm, made up of high frequencies as well as lobing, it is often better to worry more about the type of wheel specified and the dressing conditioning than the aspects of machine stability.

Cylindrical grinding is another type of grinding that is used extensively in industry. This is used often in plunge mode where the final surface finish is determined by the degree of spark-out that is allowed. This is the amount of time or number of revolutions the grinding wheel is allowed for dwelling on the workpiece. Figure 4.20 shows how the finished product, and in particular the highest peaks, depends on the spark-out. The important parameter of spark-out as far as the roughness is concerned is the speed ratio $q$, where this is the quotient of cutting speed $V_s$ and the workpiece peripheral speed $V_{ft}$ at contact.

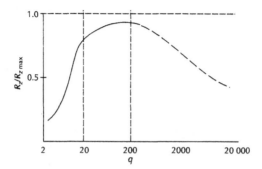

**Figure 4.20**    *Spark-out as a function of speed*

The effects of the wheel speed ratio $q$ on the workpiece roughness have to take two ranges into account: (i) the infeed per workpiece revolution is greater than the roughness (creep feed grinding); (ii) the infeed per workpiece revolution is less than the roughness (reciprocating grinding).

Note that in normal grinding, the $q$ value lies between 20 and 100 and in high-speed grinding, creep feed grinding is characterized by $q$ values of between 1000 and 20 000. Obviously, the sign of $q$ depends on whether upgrinding or downgrinding is being used (Figure 4.20).

Some approximate formulae have been derived, which relate the surface roughness value to the $q$ value at least in form [4.11]. For creep feed grinding, it has been postulated that the roughness is given by:

$$R_z = K^* |q|^{-2/3} \tag{4.6}$$

where $K^*$ involves the factors associated with grinding conditions such as the contact length and another called the 'characteristic grinding value', which loosely means the length of the grinding wheel circumference that has to grind a longitudinal element of the workpiece circumference until the value of $R_z$ has been covered in the direction of the plunge motion. For reciprocal grinding, the formula is modified somewhat to give:

$$R_z = K^* \left(1 + 1/|q|\right)^{-2/5} \tag{4.7}$$

Probably the biggest problem associated with grinding rather than simple cutting is the difficulty of dealing with deflections. What is the real depth of cut? What is the real elastic modulus of the wheel, it being composite and difficult to assess? Is the relationship between elastic modulus and hardness significant? What is true is that local elastic deformation of the wheel and the workpiece also influences the surface roughness.

Of the four components of deflection in the contact zone, three occur in or near the wheel surface while only one applies to the workpiece. The nature and constitution of the wheel therefore is most important, but this does not imply that the workpiece deflection can be totally ignored.

Workpiece hardness is an important parameter of many finished components. As the workpiece hardness increases, the surface roughness improves [4.12] owing to greater elastic recovery and the increased normal forces, which in turn produce larger grain deflection.

Wheel hardness has an effect on the surface roughness but it seems that there is some confusion as to the nature of the effect. Some workers have argued that the harder the wheel, the better the finish and vice versa [4.13, 4.14].

It seems probable, however that the real factor again is the elastic deformation length, which is determined more by the wheel elasticity that the hardness [4.15].

Depth of cut has been found to be relatively unimportant so far as the finish is concerned but wheel dressing, which might at first sight seem to be unimportant, turns out to be significant [4.16]. Poor dressing, for example, can introduce a spiral wave on the surface, which is anything but random, and can in fact dominate the roughness. Hahn and Lindsay [4.13] have shown quite clearly that if the dressing lead is very low, the surface roughness is considerably reduced, which seems obvious, or at least the other possibility is that high lead gives a high surface roughness (Figure 4.21).

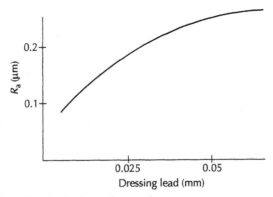

**Figure 4.21**   *Effect of dressing lead on surface roughness*

The effect of random patterned dressing passes to reduce the possibility of periodicities has not been properly explored. This technique would certainly help to target the effect of wheel dressing to its primary purpose of regenerating and sharpening the grains of the grinding wheel and not dominating the surface roughness.

The other main effect producing periodicities of the roughness – as well as the roundness as mentioned earlier – is chatter. It has already been indicated that this is due to a number of parameters of grinding, including the process machine tool and workpiece. Unlike the basic texture, chatter marks are more easily identified with specific machining parameters because of the more distinctive nature of the geometry produced.

It has been pointed out that chatter characterization is very different from that of cutting for the following reasons.

1. There is usually a slow growing rate of amplitudes.
2. Complicated waveforms are generated on the surface.
3. Grinding wheel regenerative chatter predominates.
4. There is a progressively decreasing tendency of vibration frequencies with chatter growth.

Grinding as a general process is very complicated, so it is to be expected that there is no fixed relationship between the grinding parameters and surface texture. Furthermore, because of the relatively recent growth in instrumentation for measuring roughness, many of the classical experiments have limited use because only $R_a$ or $R_z$ was measured. Nevertheless, some general results have been obtained. The implication of this will be discussed in the summary at the end of the chapter. It is hardly encouraging! The variables of importance could conceivably be listed as grinding time, wheel hardness grade, grit size and depth of cut. Some indication of their importance has been shown in the factorial experiment. For cylindrical grinding, for example [4.17], there has been an attempt to link the parameters in a non-dimensional way albeit in a very specialized experiment with alumina wheels.

It is postulated that:

$$R_a = \left( \frac{v}{V} \frac{x}{W} \frac{1}{rn_c} \frac{d}{D_e} \right)^{n/2} \varphi(a) \tag{4.8}$$

where $v$ is the workpiece speed, $V$ the wheel speed, $x$ the traverse rate, $W$ the wheel width, $d$ the effective depth of cut, $D$ the wheel diameter, $D_w$ the workpiece diameter and $n_c$ the number of active grinding grits per unit surface area of the wheel; $r$ is the average width-depth ratio of a groove produced on the surface. Values of $n_c$ vary from 1.2 to 1.4, although in high-speed grinding, the value of $n$ is approximately 3.0. The roughness is affected by the size of grain in the wheel. As a rough guide, if the grain size increases by 25% the $R_a$ value of the surface increases by about the same amount.

### 4.2.7 Nanogrinding

Traditional fine finishing processes like polishing and lapping are being replaced by **nanogrinding**. This term applies to situations where the wheel speed is high, the pressure is low and the depth of cut is of the order of nanometres. With this combination of process parameters, the cutting process finds it difficult to produce chips simply because of the energy needed to generate the surface of the chip. Also, for the same reason, any chip formed will tend to be continuous in form because to fracture the chip requires energy. This state of affairs holds even for very brittle materials like ceramics. The process under these circumstances causes 'ductile grinding'. Ductile cutting is also possible with diamond turning using light loads and high speeds.

The nanotechnology is not just that of producing an ultra-smooth finish of nanometres. It is the change in the energy balance at the tool/grain tips. The physics dictates that the energy needed to produce chips is too high: material movement has a lower energy level and will prevail. If the distance between the tool and work is kept smaller than a critical distance $d_c$, it is so small that elastic compliance cannot be tolerated. Miyashita [4.18] has produced very stiff machines to enable ductile grinding to be achieved routinely. Figure 4.22 shows a comparison between nanogrinding and conventional grinding.

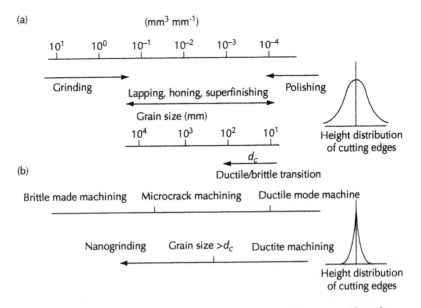

**Figure 4.22** *Conventional machining process (a) nanogrinding process (b) conventional grinding (notice the different height distribution)*

### 4.2.8   Honing and superfinishing

Honing is a finishing operation in which stones are used to provide a fine roughness of the order of 0.5μm $R_a$, often on internal cylindrical bores. The stones, which are usually made of aluminium oxide (in the form of what is called **sticks**), comprise grains of size 30-600μm joined by a vitrified or resinoid bond.

Honing is usually employed to produce an internal bearing surface for use in the cylinders of diesel or petrol engines. This involves the use of the hone as a finishing operation on the surface, which has been previously ground. For internal honing, the amount of material depth removed is small (~2μm). Traditional honing methods employ three or more stones (six is typical) held in shoes mounted on a frame that can be adjusted to conform to the correct bore size, the shoes being arranged to follow the general shape of the bore. In operation, the cutting speed is about 1ms-i with a pressure of 0.5Nmm⁻². In practice, the honing tool is allowed to dwell for a small amount of time at the end of its traverse before retraction. This causes all sorts of problems of interpretation of the surface roughness of cylinder bores because the lay changes direction rapidly.

Before more critical evaluation of honing is undertaken, it should be mentioned that another similar process is superfinishing. This is very similar to honing but the stones are given a very small axial vibration, and it is used on external cylindrical surfaces. In honing and superfinishing, the process is often used to remove the effect of abusive grinding. This is the situation in which the machining – in this case grinding – produces thermally induced stresses in the surface left by the main finishing process, grinding. A picture of a typical honed surface is shown in Figure 4.23.

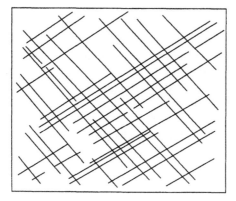

**Figure 4.23**   *Plateau honing*

### 4.2.9   Polishing (lapping)

Polishing involves the use of free grains usually suspended in a liquid rather than bonded into a wheel as in grinding. It is invariably a process associated with nominally generating ideally smooth surfaces. This process is usually defined as one in which the microroughness is commensurate with the diameter of the molecules, or of the same scale as the lattice spacing. One could argue that the smoothest surface is the surface of a liquid in an absolutely undisturbed state. A surface approaching this ideal is that of glass when congealed in a state of rest. The faces of monocrystals grown from salt solutions present another example of a surface with a fine structure [4.19, 4.20].

Of all the methods of treating a surface already mentioned, such as turning and grinding, the smoothing of the surface is at the expense of plastic deformation. The essential mechanism of fracture in plastic materials is the removal of shavings and, together with this main phenomenon, the plastic deformation of adjacent particles. However, the main phenomenon in the machining of friable materials is the development of cracks within the mass of material that penetrate to some depth below the surface and that intersect, thereby producing a mechanically weakened layer that easily fractures by the repeated action of the abrasive. This is the microfracture mode.

In polishing, there is a problem of determining the true mechanism of material removal, therefore a number of theories have been advanced. It is not the intention here to go into great detail. However, it is informative to see how one outcome can result from a variety of wholly different mechanisms. The earliest polishing theories apply to that of glass polishing. French suggested that an abrasive particle produces a pressure on the surface of glass, which has a very small resistance to fracture in comparison with its resistance to compacting, and that this causes cracks to develop along planes of maximum shear (which are inclined to the axis at an angle of about 45°). Cleavage commences along the planes of maximum shear, but due to the fact that the particles act like a wedge, the shear plane is diverted upwards, which leads to a conchoidal fracture. Preston considered the action quite differently. In his opinion, the grain rolls between the tool and the glass. Owing to the irregular shape of the grains, they produce a series of impacts one after the other, as a result of which conical cracks are generated. These cracks he called 'chatter' cracks. These intersect each other and with repeated action by the abrasive grains, pieces are pulled out. Therefore he concluded that there would be an outer layer showing relief and under it a cracked layer. The direction of the cracks and the resulting stresses in crystals differ for differently orientated faces and differ markedly from glass in, for example, lithium fluoride.

The pronounced difference in the smoothness of ground and polished glass was responsible for the theory concerning the different mechanisms of these two processes and for the development of different theories of polishing.

The real question is whether a polished surface is qualitatively different from a finely ground surface or whether it is possible to develop conditions – as in ductile grinding – under which a transition from a ground surface to a polished surface with a gradually decreasing value of roughness can be detected.

The terms 'polishing' and sometimes 'smoothing' are nominally the same as 'lapping' – the essential features being the presence of free grains in a slurry and the absence of heavy loads and high speeds. The mechanical movement required is usually by machine or by hand, in which case it can be quite haphazard.

Surface melting might, according to Bowden, contribute to the polishing process so that it could well be that the relevant physical property in polishing is not the relative hardness of the two solids but the difference in the melting points. It is interesting to note that the melting points may be a principle factor in some recent fabrication processes such as friction welding.

Above, some conventional processes have been discussed. In what follows, a few comments on unconventional methods will be made.

Figure 4.24 shows a schematic diagram, which covers most of the so-called unconventional machining processes. The main point to note is that the material removal mechanism is not confined to mechanical effects. There are many examples where thermal, chemical and/or electrical energy augments or sometimes completely replaces the mechanical contribution.

### 4.2.10  Unconventional machining [4.1]

In ultrasonic machining, an abrasive grain suspended in a slurry is used as in polishing, but there is no escape route for the particles. The machining is achieved by vibrating the tool against the workpiece. It is arranged so that the vibration node of the tool has an antinode at the tip so that the energy is at a maximum; in the other methods, the mechanical processes employ a jet of material to active machining. A water jet is very effective for thick specimens of soft material. The problem is that there has to be a tremendous pressure exerted, which requires stringent safety precautions to be in place. In this method, the water acts as its own coolant. Sand or shot **peening** is a variant in which solid particles bombard the workpiece. These methods do not necessarily remove material because the energy direction is perpendicular. Peening, in particular, is used to put compressive stresses into the surface layer of the work to reduce the risk of fatigue failure. This is used for turbine blades, for example. The actual surface finish is not the critical parameter in these methods. The removal mechanism in ultrasonic and water jet methods is by attrition. In effect, these are accelerated wear processes.

Chemical processes are very important in the semiconductor and micromechanics regimes where preferential etching can be used to produce

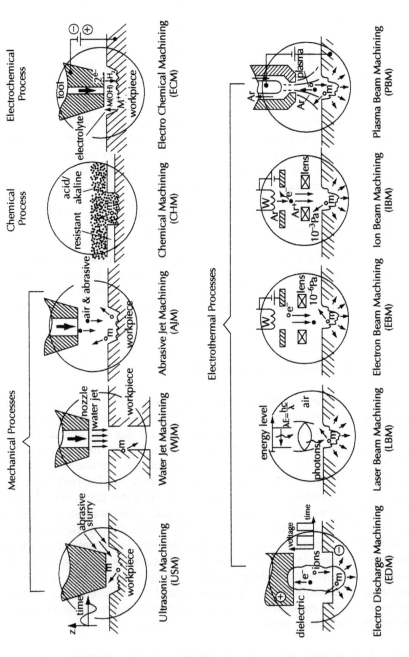

**Figure 4.24** *Mechanical and electrothermal processes [B5]*

integrated circuits to a high order of accuracy. These are 'wet' processes and rely on chemical reaction rather than mechanical energy to remove material. These methods can be used when there are very hard materials, such as ceramics, to process.

In **electrochemical machining (ECM)**, the components are immersed in the electrolyte as the anode and the material removed by means of passing an electric current through the electrolyte. This is, in effect, a non-abusive method of removal because there are no mechanisms or introduced dislocations and defects that make up work hardening. A typical schematic diagram is shown in Figure 4.25.

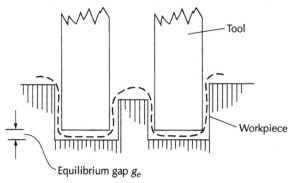

**Figure 4.25**  *Electrochemical machining*

It is possible to mix the mechanisms of the processes to get an enhanced removal at the same time as a good surface.

One example of this hybrid method is **electrolytic grinding**, which is used for very hard materials. In this, the tool is a metal wheel with abrasive grains in it. An electrolyte is circulated between the tool and the work. This gives an ECM and a grinding mechanism that can produce high removal rates in addition to surface finishes of about $0.02\mu m\ R_a$. Also, the wheel does not wear much because of the presence of the chemical etching, which takes up some of the machining.

**Electrodischarge machining (EDM)** uses the discharge of an arc between the work and the tool as the machining process. The mechanism for machining is melting, which occurs as the result of the very high temperatures (10 000°C) generated by the arc. This is well above the melting point of materials. The tool and workpiece are both immersed in oil to allow ionization to occur (Figure 4.26).

The surface roughness produced looks like cratering because of the drift of the arc. Surface roughnesses can be as small as $0.25\mu m\ R_a$ but only when the rate of machining (the sparking rate) is small.

It has already been indicated that nanotechnology requires [4.21] dimensional tolerances, form and surface finish of such small values that are almost impossible to achieve using extensions of existing technologies. A consequence of this is that

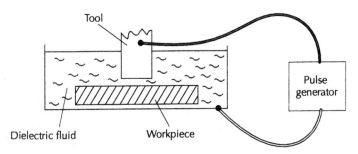

**Figure 4.26**  *Electrodischarge machining*

a number of new processes, usually based on polishing mechanisms, have been devised [4.22], mostly in Japan. Some schematics are shown below. They are largely self-explanatory and so will not be discussed here.

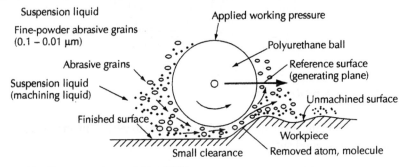

**Figure 4.27**  *Elastic emission machining (EEM)*

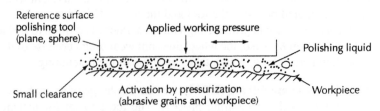

**Figure 4.28**  *Mechano-chemical machining*

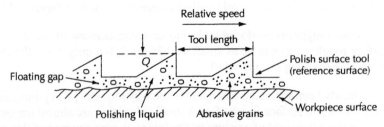

**Figure 4.29**  *Mechano-chemical machining with hydrodynamic pressure*

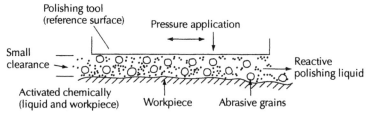

**Figure 4.30**   *Mechano-chemical polishing*

### 4.2.11 Atomic bit processes

An objective for nanomachining is to produce surfaces that are smooth to nanometres and better. Although conventional processes, such as diamond turning, can achieve these roughnesses they are not preferred. This is because the tool is large compared with the requirement.

It seems obvious that to produce surfaces that are nanometrically or even atomically smooth, it would be better to use machining methods in which the 'unit processing event' is of nanometre or atomic size. This points the way to **atomic bit processing**. In this method atoms, electrons, ions or molecules are used to do the machining. Electron beam methods have been used but tend to penetrate the surface too deeply [4.22, 4.23]. Electron beam methods are useful in lithography, in which integrated circuit masks are patterned by exposing a photoresist to the electron beam. Another preferred method uses ions to machine or sputter (Figure 4.31).

Ion beam machining has the unique property of being able to machine most non-organic materials, whether they are metals, ceramics or semiconductors. The method is that of **sputtering**, in which atoms are literally knocked off the surface. This is achieved by bombarding the surface with ions of an inert gas such as argon. These ions are accelerated to energies of about 10 k.e.V. before hitting the surface. The ions penetrate much less than in electron beam techniques. Values of about 0.1µm are typical. Very smooth surfaces (~1nm) can be achieved to smooth the surface produced by more conventional machining. However, the removal rate is very low and, even for removal of micrometres of material, hours are required. Diamonds can be sharpened, provided that charge build-up is prevented. Styli of tip dimensions of 0.01µm have been made. Usually, these methods work well for hard materials but not for softer ones. These can actually get rougher due to the wider area of energy dissipation. Figure 4.32 shows the surface roughness as a function of depth of penetration and incident angle.

As the angle from normal increases, the number of displaced atoms tends to decrease.

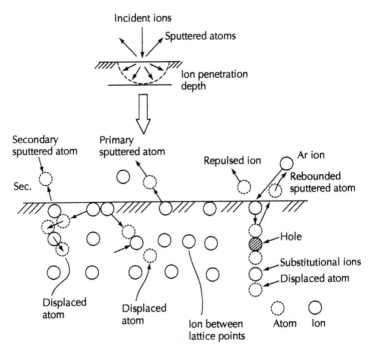

**Figure 4.31**   *Ion beam sputter – machining mechanism*

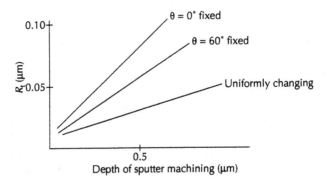

**Figure 4.32**   *Effect of changing angle during machining*

Having used a process and obtained a surface parameter, how can it be used? Section 4.3 explains how the surface can be useful to monitor the process and the machine tool.

## 4.3 Process control

### 4.3.1 Shewhart charts [4.23]

This can be achieved by using any critical parameter of the workpiece. The diameter or the texture have been used often. Control using the measured parameter is achieved by means of **Shewhart charts** or **cumulative sum (CuSum)** methods.

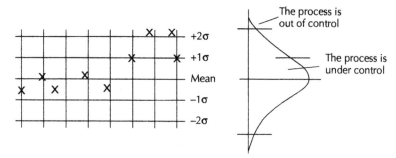

**Figure 4.33**   *The Shewhart chart*

The process is deemed to be out of control when criteria agreed between the maker of the part and the user of the part are violated in a prescribed way. Figure 4.33 shows limits of $\pm 2\sigma$. It could be $\pm 3\sigma$, if agreed. $\sigma$ is the standard deviation.

For example, each cross could be an $R_a$ value measured on the part. It could be agreed that if three consecutive measurements are outside the $\pm 2\sigma$ (95%) limits, then the machining must be stopped and remedial action taken. Perhaps the tool is broken or the wheel needs redressing. Obviously, this criterion would be meaningless if each surface parameter was subject to wide swings in value across the surface as shown in Figure 4.34.

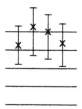

**Figure 4.34**   *Effect of excessive variations in surface parameter value*

Hence it is important to use a 'stable' parameter such as the $R_a$ value rather than the ill-behaved peak or valley or peak to valley parameters such as $R_y$. Also, whenever possible, an average reading of $R_a$ or $R_q$ should be made. If there are '$n$' readings taken over a typical area of the surface, then the spread of average readings is reduced by $\sqrt{n}$.

An alternative scheme for process control introduced by Page in [4.24] and used extensively by ICI and other large companies uses a method that shows up changes in readings that are deviating from the mean.

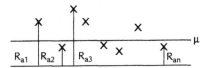

**Figure 4.35**   *Determination of deviation from intended mean value*

Consider Figure 4.35 for example. If the intended mean value of $R_a$ is, say, $\mu$ as shown, it is quite difficult from the plotted values of $R_a$ readings to determine whether or not the process average is close to the intended value $\mu$ or not.

### 4.3.2   Cumulative sums

One way around this is simply to plot the accumulation of the $R_a$ values. The method is so simple and yet elegant. It relies on a simple integral.

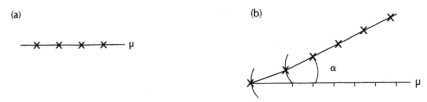

**Figure 4.36**   *(a) keeping to $\mu$ line (b) veering off $\mu$ line*

In Figure 4.36 (a), the cumulative $R_a$ is on the $\mu$ line so there is no problem, but if the cumulative $R_a$ veers off the direction of $\mu$ then there would be a problem (Figure 4.36 (b)).

If the function does not have a mean value of $\mu$ taken as zero, but some value larger at $C$ above the intended mean, the CuSum is $\int_0^L C\,dx = Cx\int_0^L = CL$ at position $L$ along the chart. The *slope* of the CuSum chart is therefore $C$. So, measuring the slope $\alpha$ gives the deviation from the intended mean value. Usually vee masks are provided, which flag a serious drift, as shown in Figure 4.37.

Should the data lie outside the arms of the vee, then there is a process problem that has to be addressed before manufacture is continued. There are two variables '$d$' and '$\theta$' (Figure 4.37), which can be adjusted to give the necessary control.

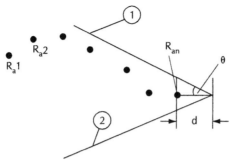

**Figure 4.37**   *CuSum techniques*

Early attempts to use the surface as a control was decided more for convenience than significance. Unfortunately, parameters that could be checked easily tended to be unreliable. Table 4.2 shows some amplitude parameters and their properties. The tendency was to control the process by the simplest possible method. Unfortunately, this approach allows serious variations.

**Table 4.2**   *Ease of measurement and reliability of amplitude parameters*

| Ease of Measurement | Reliability |
|---|---|
| $R_y$ | $R_q$ |
| $R_t$ | $R_a$ |
| $R_{tm}$ | $R_z$ |
| $R_z$ | $R_{tm}$ |
| $R_a$ | $R_t$ |
| $R_q$ | $R_y$ |

The maximum peak to valley $R_y$, for example, can be measured easily from a chart and so could be suitable for an operator-based checking scheme. $R_q$, on the other hand, is very difficult to measure off an instrument chart. $R_a$ is possible because a planimeter method is easy to use but does require more cleanliness than would be found in an average machine shop.

Spacing parameters could equally be used but tend to be more difficult to measure. Take for example the $S$ parameter – the local peak spacing. This is so dependent on the frequency response of the instrument that almost any answer could be obtained.

Table 4.3 shows the possibilities for checking process and machine tool properties.

**Table 4.3**   *Checking processes and machine tool properties*

| Property | Parameter |
|---|---|
| Depth of cut | $R_{tm}, R_y$ |
| Feed | $S_m$, HSC |
| Material removal | $R_a, R_q$ |
| Process stability | Variability of parameters |
| Tool wear | Power spectrum |
| Wheel dressing | Autocorrelation |
| Micro fracture | $S, \Delta_a$ |
| Long term instability | Space frequency functions |
| Chatter | Power spectrum lobing |
| Axis of rotation | Roundness |
| Slideway/load screw | Straightness |
| Clamping | Lobing |

## 4.4 Relationship between surface metrology and manufacture

Figure 4.38 shows the trend, which is to move the measurement to where it is most needed i.e. at the function. If the function is understood and the process is controlled, no metrology would be required as shown in (d). Figure 4.38 (a) shows the old state of control (or lack of it); the part is made and tried out. In 4.38 (b), measurements are made on the specimen after it has been made. These are then plotted on a Shewhart chart, cusum chart or any other control method. Figure 4.38 (c) is basically a 'goods inwards' check to ensure that the 'as made' part will perform well. Figure 4.38 (d) is ideal. There is so much 'capability' in the system that no measurement is needed because the link-up between manufacture and use has been properly made. Needless to say, 4.38 (d) is rarely achieved, but if it is, checks have to be made periodically.

Figure 4.39 sums up the situation in terms of the surface. It shows the general system, which splits into three regions: A, where the manufacture is controlled, B, where condition monitoring and surface integrity can be checked by examination of the surface, and C, which is concerned with the use of the workpiece and how the surface is important.

The essential point is to determine the objective at the start of any attempt to correlate the process and machine tool with the surface. The first and most important point is that using the surface is not a second best option. It is *at the surface* where the energy transfer from machine tool to chip takes place. All other

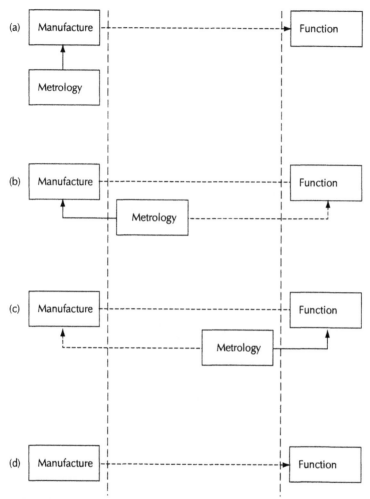

**Figure 4.38**   *Relationship between manufacture and surface metrology*

measuring systems such as force or ultrasonics are more indirect. The only problem with surface measurement is that it is quite difficult to achieve in real time, although in situ methods are readily available. These options will be considered in the section on instruments.

It can be seen from Figure 4.40 that there are two loops in the conventional machine tool. One is the force loop, which tries to withstand the inertial energy from the chip from the workpiece and at the same time deflect the tool column. The other is the metrology loop, which may sometimes be coincident with the force loop. It is invariably open loop. It does not close on the workpiece and it has no feedback.

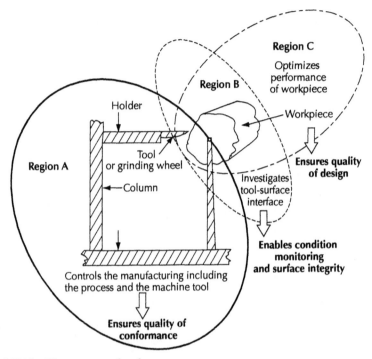

**Figure 4.39 (a)** *Why measure surfaces?*

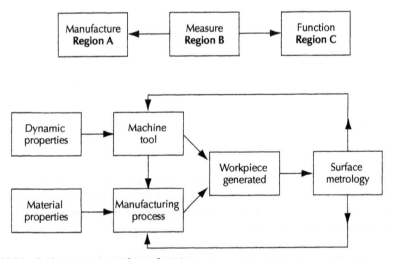

**Figure 4.39 (b)** *Surface measuring and manufacturing*

All aspects of manufacture can be helped by examining the surface finish, as seen in Figure 4.39, in block schematic form and Figure 4.40 as a cartoon.

There are two basic loops (Figure 4.40):

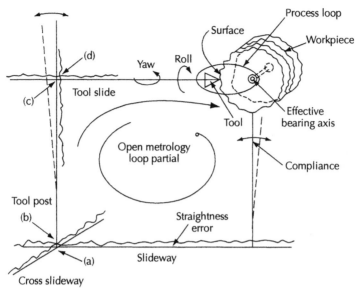

**Figure 4.40**    *General system*

1.    Machine tool loop → often containing the metrology system – this is usually open loop.
2.    Process loop → closed on interface between chip and surface.

The surface is where the loop is closed and control can start.

Also shown at the surface is the process loop, which in effect takes in the bearing of the rotating part (or wheel in grinding).

## 4.5 Force and metrology loops

The force and metrology loops need clarifying. The force loop provides load on the workpiece and generates the motion of the workpiece. In a completely static system, the force loop is closed but in a dynamic system it can be momentarily open loop. The metrology loop is always in the form of a calliper as mentioned earlier, one arm touching the test piece and the other the reference. The metrology loop carries information rather than energy or force.

Because of these two different requirements, the two loops should be kept apart. Unfortunately, slideways-carrying metrology scales also carry load. At the surface, the opportunity arises to measure with little interference from the force loop. For example, curvature and hence the radius of the workpiece can be evaluated from the surface itself and does not have to rely on radius measurement generated by the tool arm. This is because it is in the force loop causing deflections at the workpiece and violating calibration.

Curvature measurement i.e. $\dfrac{d^2r}{d\theta^2}$ is the reciprocal of the radius. Any errors in curvature are fed back via the short closed loop containing the yoke to the tool position.

What is not realized is that the surface can act as its own reference within pre-scribed limits (discussed in the section on instrumentation). Figure 4.41 shows a typical system. The yoke connecting arms (1) and (2) provides the reference arm to the transducer whereas probe (3) is the test arm. Here, the surface provides the reference arm and the test arm.

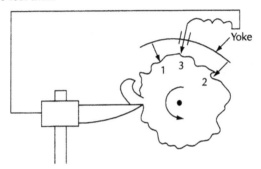

**Figure 4.41**   *Intrinsic reference*

In this case, the metrology loop provides the demand signal for the force member carrying the tool. In addition, there is the energy consideration. The stable situation lies where the surface free energy of the chip and the surface roughness is a minimum. This urge to reduce energy is one of the internal constraints on the cutting. There is, in effect, a natural law that is controlling the roughness. Energy minimization at the chip-tool surface is an instant control, thereby acting as a tightly controlled closed loop at the point where the tool and chip intersect.

Roughness values should be taken from the workpiece on the machine tool but this cannot be achieved while the specimen is being machined.

## 4.6   Unit events and autocorrelation [4.1]

The objective is to optimize the process. This cannot be done by using the simple parameters e.g. $R_a$, which can control the process but not optimize it. However, there is a possible way to optimize using the autocorrelation and power spectrum. Pilot tests show how this is done. The optimization is based on examining the unit event of machining. In grinding, this is the effect one grain has on the surface .

Take grinding, for example, as a case where autocorrelation is useful. Figure 4.42 (a) shows the impression left on the surface by a sharp grain. Figure 4.42 (b) shows a typical profile and Figure 4.42 (c) the correlation function. Notice that the

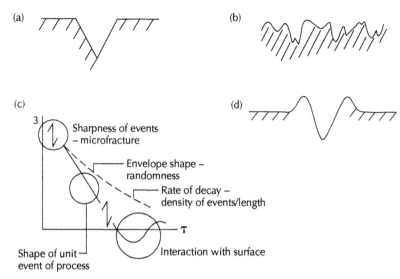

**Figure 4.42**  *Autocorrelation and grinding*

correlation length (distance over which the correlation drops to nearly zero) is a direct measure of the effective grain hit width. For a grinding wheel in which the grains are blunt (Figure 4.42 (d)), there is a considerable material pile-up as well as a chip formed. By examining the correlation function, it is apparent that pile-up or ploughing is revealed by a lobing shape in the autocorrelation function. The width of the central lobe is a measure of material removed. At a glance, therefore, the shape of the autocorrelation function reveals the efficiency of the grinding in all its aspects (Figure 4.42). Notice that this would not be revealed by looking at the profile or by using simple parameters. In Figure 4.42 (c) any longer waves in the autocorrelation function of the surface show other problems such as the need to dress the wheel.

## 4.7   Use of the power spectrum [4.1]

Another example shows how the power spectrum can be used to identify problems in turning. As the tool wears and the machine tool deteriorates, significant changes occur in the spectrum of the surface as shown in Figures 4.43 and 4.44.

Figure 4.43 (a) shows a profile of good turning produced by a good tool; Figure 4.44 (a), together with its spectrum, shows some line frequencies, the fundamental corresponding to the feed and a few harmonics due to the fact that the shape of the tool is not sinusoidal.

As the tool wears, the ratio of the harmonic amplitudes to that of the fundamental increases (Figure 4.43 (c)). This is due to the imposition on the surface of the wear

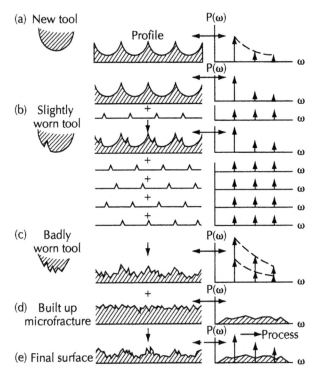

**Figure 4.43**   *Power analysis harmonics*

scars on the tool. Also on this right hand side of the fundamental spectrum, the base line can rise, due to random effects of the chip formation and micro-fracture of the surface.

Figure 4.44 shows another factor i.e. how the machine tool can be monitored. Take, for example, turning as before in Figure 4.43. This time, however, look at the longer wavelengths of the surface. Because they are bigger than the feed, they appear as sub-harmonics in the power spectrum. Because of this, some components (i.e. arrows) are present between the origin and the feed frequency i.e. region A so that by using random processes to analyse the surface it is possible to investigate the process and the machine tool in a way not possible by any other method. Remember that the output is a direct measure of performance – it is the workpiece.

## 4.8   Application of space frequency functions [4.25]

From what has been said above, it could be argued that autocorrelation and power spectrum are sufficiently comprehensive to be able to cater for most eventualities in manufacture. Unfortunately, this is not so. There are instances where subtle changes in the surface geometry can be very important in machine tool

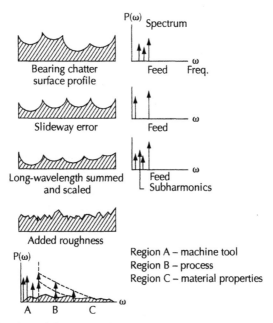

**Figure 4.44**   *Spectral analysis sub-harmonic and total spectrum*

monitoring. For example, the mode of vibration of a tool column determines the changes in statistics of the surface geometry. The power spectrum or autocorrelation cannot detect changes in the statistics. Unfortunately, changes in the nature of the surface often accompany the presence of defects, flaws etc. that are detrimental to the performance of the workpiece.

Looking at the formulae for autocorrelation $A(\tau)$ or power spectrum $P(w)$:

$$A(\tau) = \frac{1}{l} \int_{\frac{L}{2}}^{\frac{L}{2}} f(x) f^*(x + \tau) dx \tag{4.9}$$

$$P(w) = \left| \int_{0}^{\infty} f(x) \exp(-jwx) dx^2 \right| \tag{4.10}$$

These show that the integral limits extend over all the signal $f(x)$. The whole signal is integrated. Any change within these limits in the nature of $f(x)$ will simply be averaged out.

This restrictive behaviour of the random process functions has been recognized for some time and measures have been taken to remedy the situation by modifying the definitions. In effect, there has to be a window function introduced in either time or frequency, which would localize the extent of the signal $f(-)$ being examined. The width and shape of the window function, together with its position along the time or frequency axes, provide adequate flexibility and pinpoint examination of small detail in the signal. With this incorporation, the 'time frequency' functions have evolved. In surface metrology, the time variable is replaced by the spatial dimension $x$.

There are many forms of 'space-frequency' functions. These include Wigner, ambiguity, Gabor functions as well as wavelet functions. In the first two, the function inside the integral sign is modified.

Although the potential of these complicated functions is recognized, their use has only been rudimentary because most production engineers and instrument engineers are not familiar with them. This state of affairs is changing with more widespread availability of research papers [4.25]. One such application follows. This is concerned with the machine tool rather than the process.

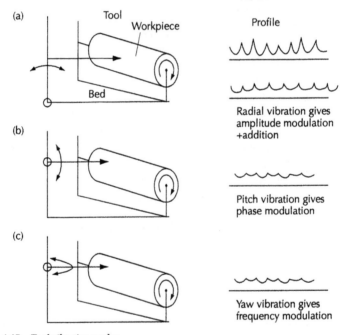

**Figure 4.45**   *Tool vibration modes*

All of the vibration modes shown in Figure 4.45 are difficult to analyse using conventional random process analysis. Not included above is tool roll, which is even more difficult to analyse. Wigner distribution analysis easily separates them [4.25].

Wigner distributions tend to be better at identifying the nature of the modulation ambiguity where it occurs than wavelet transforms, which are good at identifying flaws and defects rather than changes in the statistics.

It could be argued that all these functions are too complicated to be used in what is, in effect, a practical subject. This is not so; modern machine tools are complicated but they are in general use. What is true for hardware is true for software!

It is clear from above that formidable analytical tools can be used to extract manufacturing information from the surface. This means that the surface can be used in machine tool diagnostics as well as in process control. However, neither of these applications is the important one. The fundamental reason for making the workpiece is to use it. It is in the use of the workpiece where the surface is most important and this will be explored next.

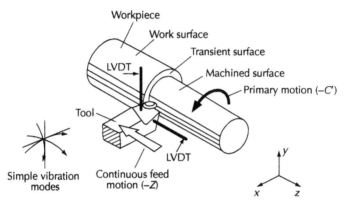

**Figure 4.46**   *Domain of cutting mechanism*

Pitch, roll, and yaw all produce different effects on the surface. Conventional parameters find it difficult to separate these degrees of freedom.

## 4.9   Conclusions

The surface can be used to progress all aspects of quality in manufacture. Sometimes, very simple parameters such as $R_a$ or $R_z$ can be used to control the process such that changes can be detected quickly and effectively. The surface of a workpiece is readily available and can be measured either close to it or removed from it. Some of the more complicated parameters need to be used to monitor and prevent machine tool problems from developing. The point is that these techniques can now be used. Whether they are or not depends on being able to implement them without getting an adverse reaction due to unfamiliarity. Alternatively, they may not be used because they are not commercially available on instruments. Whatever the reason for not using these new techniques, there is no excuse for not being aware of their potential.

## References

4.1  Whitehouse D. J. *Handbook of Surface Metrology*. IOP Pub. Bristol (1994).

4.2  Opitz H and Mall H. *1-lerstellung Hochwertizwer brehflachen Ber Betriebswiss* (Berlin: VDI) (1941).

4.3  Bramertz P. H. *Ind. ANZ* Vol. 83 p525 (1961).

4.4  Hingle H. *Control of Precision Diamond Turned Surfaces*. Proc. VIII. IN. Oberflachen. Kolloq Chemnitz (1992).

4.5  Radford J. D. and Richardson D. B. *1974 Production Engineering Technology London*. Macmillan (1974).

4.6  Dickenson G. R. Proc. I. Mech. Eng. Vol. 182 p135 (1967/68).

4.7  Tlusty J. Zaton W. and Ismail. Annals CIRP Vol. 32 p309 (1983).

4.8  Martelloti M. E. Trans. AME Vol. 63 p677 (1941).

4.9  Salje E. *Relations between Abrasive Processes*. Annals CIRP Vol. 37 p641 (1988).

4.10 Bhateja C. R. Annals CIRP Vol. 33 p199 (1984).

4.11 Salje E. Teiwas H. and Heidenfelder H. Annals CIRP Vol. 32 p241 (1983).

4.12 Saine D. P. Wagner J. G. and Brown R. H. Annals CIRP Vol. 31 p241 (1982).

4.13 Hahn R. S. Lindsay R. P. Annals CIRP Vol. 14 p47 (1966).

4.14 Hahn R. S. Trans. ASME Vol. 186 p287 (1964).

4.15 Konig W. and Lrotz W. Annals CIRP Vol. 24 p231 (1975).

4.16 Pahlitzch G. and Cuntze F. O. Private Communication (1980).

4.17 Loladze T. N. Annals CIRP Vol. 31 p205 (1982).

4.18 Miyashita M. and Yoshioka J. Bull. Jap. Soc. Prec. Eng. Vol. 16 p43 (1982).

4.19 French J. W. Trans. Opt. Soc. Vol. 18 p8 (1917).

4.20 Preston F. W. J. Soc. Glass Technol. Vol. 17 p5 (1983).

4.21 Miyamoto J. and Taniguchi N. Bull. Jap. Soc. Prec. Eng. Vol. 17 p195 (1983).

4.22 Taniguchi N. Annals CIRP Vol. 132 p573 (1983).

4.23 Shewhert W. A. *Economic Control of Quality of Manufacturing Products*. Nan Nostrand, Princeton (1931).

4.24 Page E. S. *Continuous Inspection Schemes*. Brometrika Vol. 41 p150 (1954).

4.25 Whitehouse D. J. and Zheng K. G. *The Use of Dual Space Frequency Functions in Machine Tool Monitoring* J. Inst. Phys. Meas. Sci. and Tech.Vol. 3 p796 (1992).

# 5

---

# Function and surface texture

## 5.1  Generic approach

Relating surface texture to performance, here called **function**, has always been difficult. This is because the key ingredients of tribology, surface instrumentation, random process analysis and digital analysis all evolved at about the same time in the late 1950s and throughout the 1960s. As a consequence of this simultaneous development, some good instrumentation was not backed up properly with good theory. The development has been uneven and has resulted in a lack of structure.

Probably the biggest problem has been a lack of focus when discussing function. This has been due to two reasons. The first is the fact that the number of applications where surfaces are important is so large and diverse that a systematic approach has been impossible. The second is that in many cases where a relationship between the surface and function has been found, the link has been suppressed due to confidentiality. For these reasons, a generic approach is given here to provide a guide. A major aspect of this approach is the classification of the function into a simple format. Obviously, no classification can cover all functions but the approach here has been to classify tribological applications and in particular, contact, friction, wear, lubrication and failure mechanisms. These make up the bulk of engineering functions.

The characterization of function, as indicated earlier, has been broken down into two variables.

1.  The normal separation of two surfaces.
2.  The lateral relative velocity of the two surfaces. Some of the tribological functions are indicated in Figure 5.1.

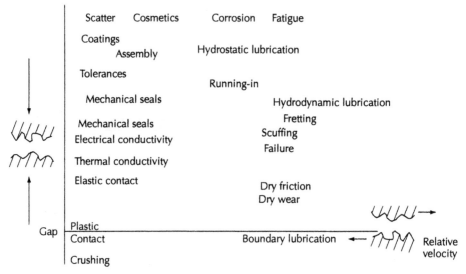

**Figure 5.1**   *Function map*

The function map treats the gap between the surfaces and the relative velocity as *variables*. This enables some *measure* of performance to be made. It is, however, possible to treat them as *attributes*. An example is 1996 ISO 1208 Annexe B p 828, which attempts to link motif characterization with function. The branching is shown in Figure 5.2.

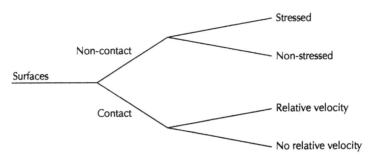

**Figure 5.2**   *Surface classification – by attributes*

This map in Figure 5.1 is here called a 'function map'. What is omitted from this diagram are the scales. In general terms, the ordinate axis is in micrometres. Elastic contact, for example, could be 0.1μm as shown in Figure 5.1. Also, the abscissa could have a maximum realistic value of 5m/sec. Not all of these functions are subject to the same forces. A 'force map' is shown in Figure 5.3. The designer should be aware of these different force regimes when specifying function for surfaces.

It is interesting to note that the force map fits comfortably into a second order dynamic system – indicated by the curve . Notice that this force map is concerned with **gap properties** and not with individual surfaces. The gap can be liquid or air filled and have laminar or turbulent flow i.e. moving from inertial to viscous forces.

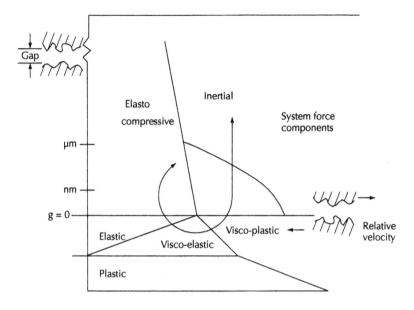

**Figure 5.3**   *Force regime map*

The objective is to use Figures 5.1 and 5.3 to fit surface characteristics. Before this, it should be remembered that the geometric properties of the surface can aid or inhibit the functional properties.

A simple breakdown of this aspect is shown in Figure 5.4.

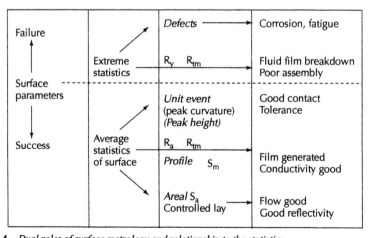

**Figure 5.4**   *Dual roles of surface metrology and relationship to the statistics*

The general rule from this figure and Figure 5.1 is that *average* geometrical properties of the surfaces (and hence the gap) improve performance, whereas extreme values tend to ruin performance. An average statistic could well be the $R_a$ and an extreme could be $R_{max}$ or $R_y$.

The 'defect' category does not mean an extreme of a distribution; it means the presence of unusual features on the surface not at all concerned with surface statistics. This category is called 'singular'. Table 5.1 lists some examples.

**Table 5.1**   *Types of surface parameter*

| Averages | | Extremes | Singular |
|---|---|---|---|
| Heights | $R_a$ $R_q$ | $R_t$ $R_{max}$ $R_{tm}$ | Point defects Scars |
| MR(%) curve | | $R_z$ | |
| Spacings | HSC $S$ $S_m$ $\lambda_a$ | | |
| Hybrid | $\Delta_q$ | | |
| Same statistical distribution | | | Different statistics |

To put the surface metrology in perspective relative to the function map of Figure 5.1, consider Figure 5.5. This breaks the map more or less into 'layers of metrology' starting at dimension and moving down the figure.

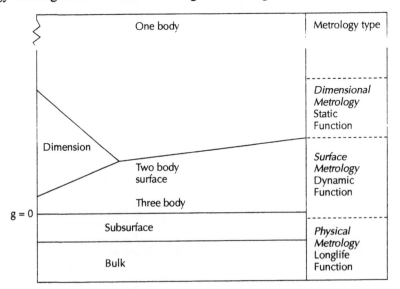

**Figure 5.5**   *Application of metrology chain*

Figure 5.5 shows the chain of metrology measurements as a function of the gap and velocity. The sequence has to start at the top and proceed downwards to the bottom of the graph. This corresponds to the gap axis of Figure 5.1. In terms of the metrology, the order of inspection is (1) dimension (2) surface (3) physical.

Before placing surface parameters into the function map, it is important to realize that within one function, more than one aspect of surfaces (gap) may be important. Also, the function itself has to be examined.

The metrology chain shown in Figure 5.5 shows the order of inspection for conventional size components i.e. from dimension through surface and subsurface. For these, it is relatively easy to separate dimensional effects in tolerancing from that of the surface because they are orders of magnitude different. However, for small parts < 1mm, roughness and dimension tend to be indistinguishable so the inspection of both characteristics has somehow to be integrated. An example of successful integration is the measurement of roughness and radius using a stylus method that has a wide transducer range as well as high resolution. Examples of integrated measurements are shown in Chapter 6.

Consider, for example, contact phenomena. There are two factors making up the function. One is the 'unit event' of the function. This is the name given to the typical mechanism i.e. the average contact is called the **unit event** here . The other consideration is the distribution of these 'contacts' in space. The surface geometry is obviously important in determining the contact characteristics. For example, the curvature of an asperity determines the elastic deformation.

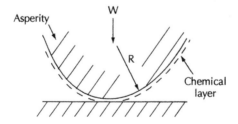

**Figure 5.6**   *Effect of chemical film on contact*

The radius of contact '$a$' is given by $a = K\left(\dfrac{WR}{E}\right)^{\frac{1}{3}}$, where $K$ is a constant ($\sim1.1$) and

$W$ is the load. The value $R$ is the extent of the surface involvement. There are other factors that have to be taken into account such as $E$, the elastic modulus. Here, $E$ is a composite value, where $\dfrac{1}{E} = \dfrac{1}{E_1} + \dfrac{1}{E_2}$, $E_1$ and $E_2$ being the respective elastic moduli of the two surfaces making contact. This is the situation for simple contact.

There are occasions in contact where there are other important factors – in electrical and thermal conductivity for example. If there is any film on the asperity such as an oxide or sulphide, it will dominate the contact behaviour (see Figure 5.6). Under these circumstances, the role of the surface geometry is reduced. It may be 30% important. However, the combined geometry of the two surfaces dominates how and where the contacts will occur. The geometry more or less completely dominates the distribution of contacts. It can be seen therefore that surface geometry, taken as a whole, is probably the most important parameter for contact, whether surface films exist or not.

This proposition has been realized for some time but even modelling the surface with the two factors unit contact and contact distribution, has proved to be difficult.

Earlier models for surfaces developed in the 1960s comprised of fixed radius hemispheres representing the peaks. These were distributed on the surface such that their height distribution was Gaussian in character. These types of model are a mixture of deterministic detail: the shape and radius of the peaks, and a random distribution of heights.

Later, in the 1970s and 1980s, the models became more realistic when it was realized that random process theory could adequately describe peaks having a range of curvatures as well as a random distribution of heights. So, using this scheme, the models were consistent.

Figure 5.7 shows the units of the function map. Both the gap between the surfaces and their relative velocity are treated as variables.

Figure 5.8 gives a very general idea of the distribution of averages, extremes and defects on the surface with relation to the function.

This should be compared with Tables 5.2 and 5.3, which are other attempts to relate function and surface. Notice the reluctance to treat the surface parameters as variables i.e. the specific value is given. Everyone regards the parameters as attributes i.e. is it important or not? Tables 5.2 and 5.3 have what at first sight seems to be more information than Figure 5.1 but in fact have not. Attribute tables reflect the lack of real information!

## 5.2 Some specific examples in tribology

(1) *Contact.* As described earlier, the average parameter so far as the unit event is concerned is the mean radius of curvature of the peaks. Important from the distribution point of view are the highest peaks. These determine the position of first contact, which can be important in electrical conductivity.

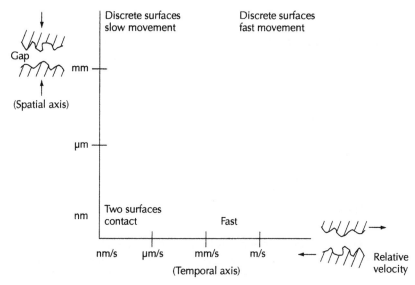

**Figure 5.7**  *Scales of size of variables*

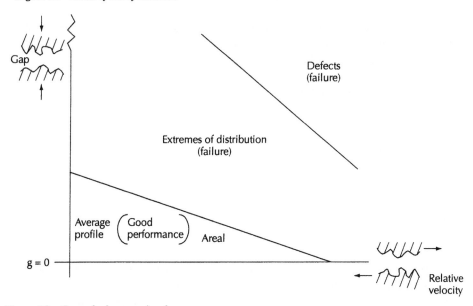

**Figure 5.8**  *General relevance of surfaces*

(2) *Lubrication.* In the first instance, consider hydrodynamic lubrication. Here, a fluid film has to be set up by virtue of the motion in order to carry the load of the shaft. There are two influences of the surface geometry. The first case is that of high peaks on either surface. These can cause metal to metal contact at high speed, which can cause film failure. This can be due to the heating of the liquid by contact. This in turn reduces the viscosity, which then reduces the film and thereby

creates more contact until a 'thermal runaway' occurs, which causes a failure e.g. in scuffing. If the surface is reasonably uniform, the areal surface texture can be important. Figures 5.9 (a) and (b) show two configurations.

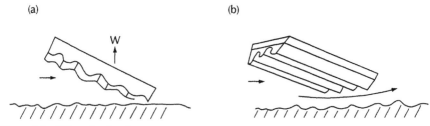

**Figure 5.9**   *Effect of surface lay*

In Figure 5.9 (a) the roughness opposes the movement, thereby increasing the friction and improving the load-carrying capacity of the bearing pad. In case (b), the surface texture is in the same direction as the flow, thereby helping it. This implies a lower coefficient of friction but conversely a poorer load-carrying capacity.

The way in which the two figures are drawn suggests that in Figure 5.9 (a), roughness increases lift off, whereas going along the lay i.e. in the tool path reduces frictional drag in the direction of flow.

A general observation follows.

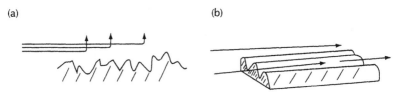

**Figure 5.10**   *Load carrying capacity vs frictional coefficient*

1.   The process marks as shown in Figure 5.10 (a) have an effect at right angles to the plane containing the roughness. In the function map approach, the process roughness tries to widen the gap 'g' i.e. the load-carrying capacity increases but the friction is high.
2.   The machine tool marks as in Figure 5.10 (b) have an effect in line with the flow i.e. tries to increase the flow and therefore reduce the friction. This situation is shown in Figure 5.11.

So, in this case the manufacturing effects can overlay the function map directly as shown in Figures 5.11 and 5.12. Notice the extreme simplicity.

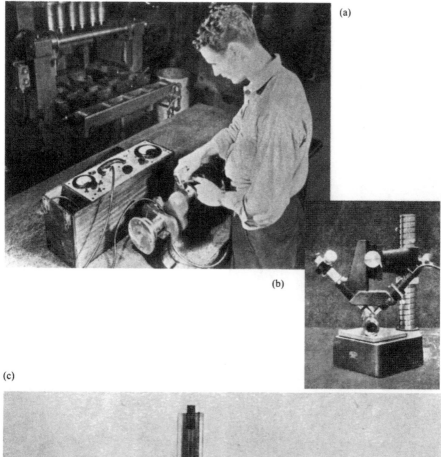

**Plate I** *Some early instruments (a) Abbotts Profilometer (b) Schmalz photomicroscope (c) Taylor Hobson surfacemeter (before Talysurf name) (courtesy Taylor Hobson Ltd)*

(a)

(b)

(c)

(d)

**Plate II** *(a) Hand-held instrument (b) Sutronic Duo instrument measuring surface of cylinder head and piston (c) Remote operation – numerical output (d) Remote operation (courtesy Taylor Hobson Ltd)*

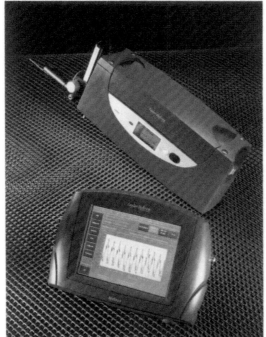

**Plate III**   *Talysurf Intra – shows display (courtesy Taylor Hobson Ltd)*

(a)

(b)

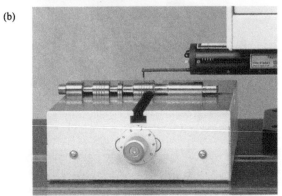

(c)

**Plate IV**  *P.G.I. Integrated instument (a) Interchangeable stylus (b) Wide range (c) Versatile positioning (courtesy Taylor Hobson Ltd)*

(a)

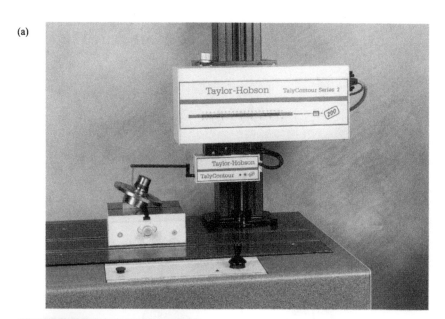

(b)

(c)

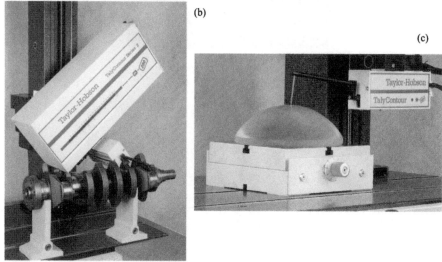

**Plate V**   *TalyContour Series (a) and (b) Show calibration (c) Demonstrates wide range, high resolution (courtesy Taylor Hobson Ltd)*

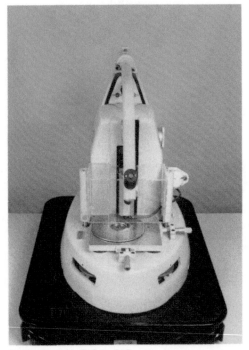

**Plate VI**   *The Talystep (courtesy Taylor Hobson Ltd)*

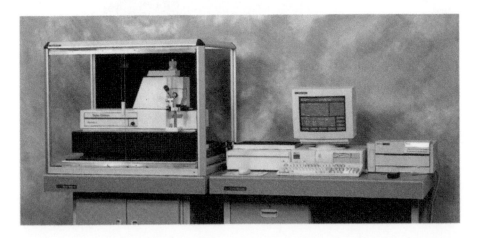

**Plate VII**   *The Nanostep, successor to Talystep, illustrates environmental insulation (courtesy Taylor Hobson Ltd)*

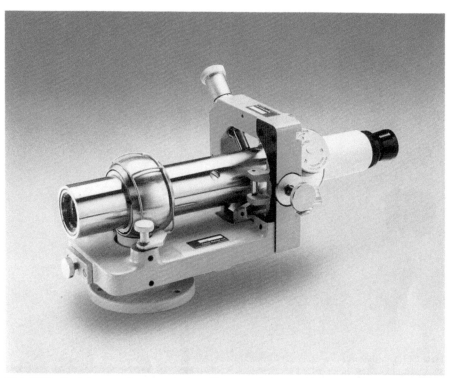

**Plate VIII**   *Straightness – alignment telescope. Illustrates pivot around spherical collar (courtesy Taylor Hobson Ltd)*

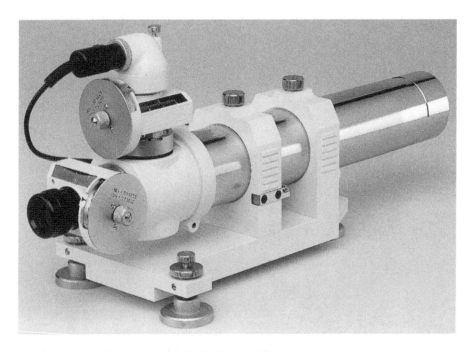

**Plate IX (a)**   *Autocollimator (courtesy Taylor Hobson Ltd)*

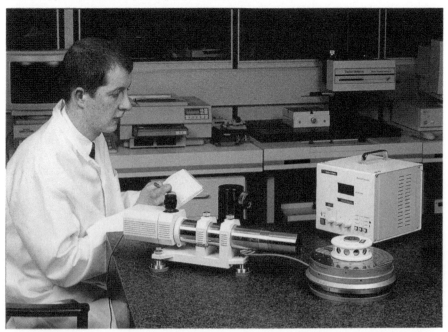

**Plate IX (b)**   *DA20 Autocollimator with reference index table used to calibrate a polygon (courtesy Taylor Hobson Ltd)*

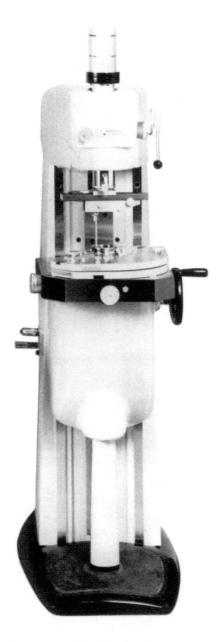

**Plate X (a)**  *A rotating spindle instrument Talyrond 73. The ultimate in rotational accuracy (courtesy Taylor Hobson Ltd)*

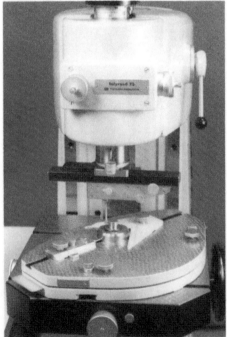

**Plate X (b)**   *Internal cylinder measurement, and external illustrates workpiece positioning methods (courtesy Taylor Hobson Ltd)*

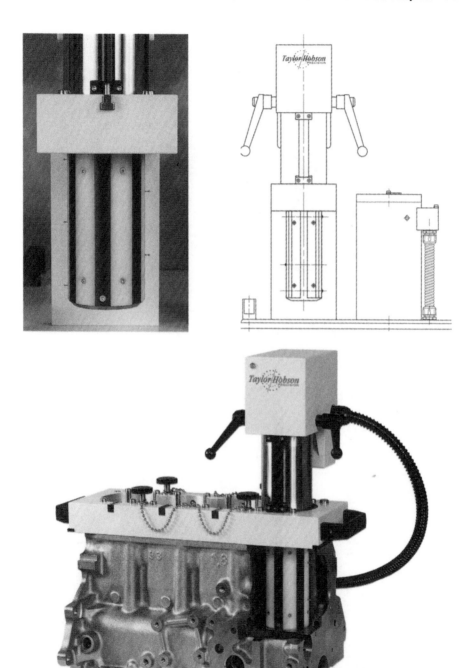

**Plate XI**  *Portable rotating pick-up roundness measurement shows the measurement of cylinder bores (courtesy Taylor Hobson Ltd)*

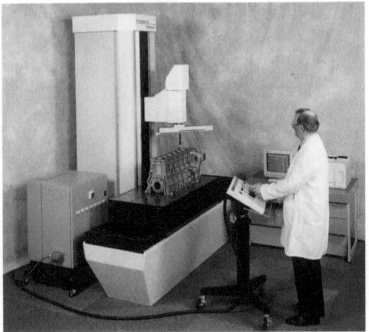

**Plate XII**   *Large rotating spindle roundness instruments (courtesy Taylor Hobson Ltd)*

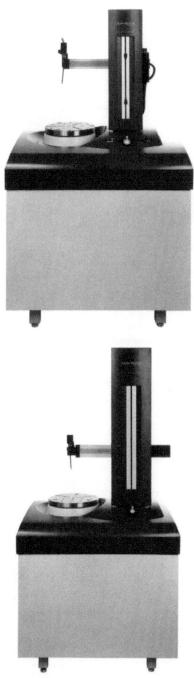

**Plate XIII** *Talyrond 265 series – workshop and laboratory. Illustrates modularity, in this case of column (courtesy Taylor Hobson Ltd)*

**Plate XIV**    *Talyrond 30 – roundness measurement in its simplest form (courtesy Taylor Hobson Ltd)*

**Plate XV**    *Talyrond 31 – cylindricity measurement (courtesy Taylor Hobson Ltd)*

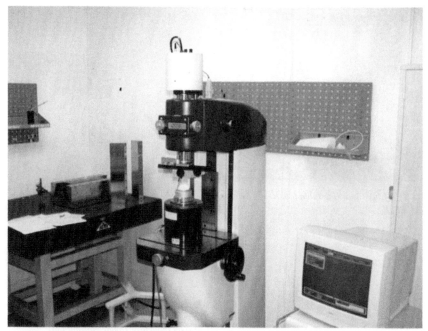

**Plate XVI**   *Set-up apparatus – roundness calibration (courtesy Taylor Hobson Ltd)*

**Plate XVII**   *Calibration and curvature artifact (courtesy Taylor Hobson Ltd)*

**Plate XVIII (a)**   *Indexing table (courtesy Taylor Hobson Ltd)*

**Plate XVIII (b)**   *Taylor Hobson Centre of Excellence NAMAS Roundness Calibration System (courtesy Taylor Hobson Ltd)*

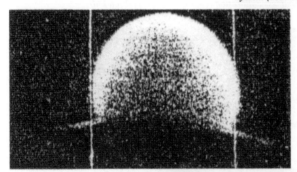

*Reflow PSG hemisphere*

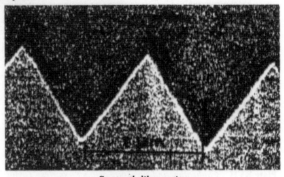

*Sawtooth-like grating*

**Plate XIX**   *Nanoscale calibration artifacts (courtesy Taylor Hobson Ltd)*

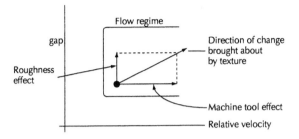

**Figure 5.11**  *Effect of texture on flow*

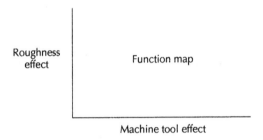

**Figure 5.12**  *The function map and manufacture*

Note that for the same amplitude of waveform, whether for process roughness or machine tool effect, the pressure gradient in the direction of flow is given by:

$$\left.\begin{array}{l} \dfrac{dp}{dx} \sim \text{Reynolds Equation} - f_{MT}(x) \\[2mm] \dfrac{dp}{dx} \sim \text{Reynolds Equation} + f_{p}(x) \end{array}\right\}$$

$f_{MT}(x)$ is machine tool error effects, $f_p(x)$ is process roughness.

3.    *Running-in – shakedown*. For many years, engines have been worked over a period of time to allow a settling down period to smooth over any small engineering mismatch. This period is called the 'running-in'. After this time, it is assumed that little will change over a long lifetime.

**Figure 5.13**  *Effect of running-in*

Figure 5.13 shows a surface of a car cylinder liner before and after run-in. Two functional requirements have to be satisfied. The first is that all deformations due to loading will be elastic, not plastic, and the other is that oil will be retained on the wall of the cylinder at all times. There is, in fact, a third, which is concerned with wear. As the running-in progresses, the hardness of contacting members will converge so that there is no big difference in hardness between the piston and cylinder. This is a result of work hardening. So, examination of the peaks on the run-in liner show that their curvature will be considerably reduced from the original values. Under these conditions, not only is the average peak curvature under the elastic limit but so also are the small local values of curvature.

This is a good example of a 'designer surface'. Automotive engineers have arranged that the surface texture of the cylinder liner is machined to be the run-in surface at the start of the engine life. The idea is to simulate the run-in process. This philosophy has worked. The run-in period has been reduced from 50,000 km to approximately 500 km. The final surface is achieved by means of a two process approach. To start with, the liner is rough honed to get the deep valleys required for oil retention and then fine honed or ground to get the load-carrying capacity under elastic conditions. This double process is called 'plateau honing'.

4.   *Corrosion*. Corrosion centres can easily be generated in deep valleys. These progress to damage the substructure of the surface. Table 5.3 shows current thinking about the influence of the surface.

5.   *Tolerances*. Vectorial tolerancing now makes the surface much more important. The global old-fashioned tolerancing, which took no account of the most sensitive positions and directions, was grossly inefficient. Now, the position as well as the values of extremes will be required. See Figure 5.14.

It is fortunate that vectorial tolerance is becoming used because the interaction of surface roughness with dimensional tolerance will get larger as products get smaller. At less than 1mm$^3$, it is becoming difficult to keep surface and dimensional metrology apart!

Process control using dimensional analysis has, in the past, been loosely targeted – current thinking requires a better understanding of the implications of poor dimensional control and in particular how a better understanding of tolerancing can ease the need for expensive machining – where is the fine tolerance really needed?

The old blanket approach to tolerance is insufficient to control the process – the information bandwidth is too low! Vectorial tolerancing allows sending details of deviations from CAD to CAM and CMM with the same format. The measured deviations are sent from the CMM to CAM by means of CAD.

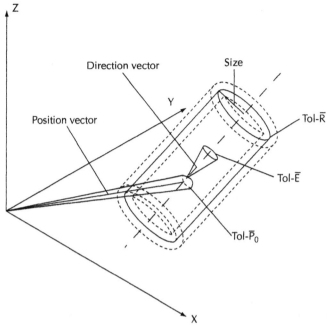

**Figure 5.14**  *Vectorial tolerancing for a cylinder*

## 5.3  Surface models

So far, it has been seen that surface models have been based on communication theory. The adoption of this theory was a breakthrough in terms of describing surfaces consistently.

From early sections on manufacture, it is evident that using random process analysis to support the conventional surface parameters has enabled the use of surface analysis to be extended from SPC to machine tool monitoring. Therefore, so far as manufacturing is concerned, the theory is adequate. For example, if only one surface needs to be considered, there is no need for 'gap' or 'lateral movement' except in descriptions of tool position.

However, current thinking indicates that even this model needs modifying to better reflect the mechanical situation that occurs, for example, in contact where the surfaces contact top-down on each other. Another problem that has to be considered is that contact occurs in a parallel mode rather than a serial one. The conclusion has been reached that random process analysis is adequate for monitoring the manufacturing process, but is inadequate for some cases of functional prediction.

The reason is that the foregoing theory was developed for communication and not for tribology. Take, for example, the simple case of loading a peak, as shown in

Figure 5.15. Random process analysis can give an idea of the density of peaks and, to some extent, the heights, but cannot as a rule determine which peaks deform under load.

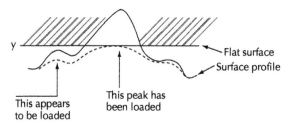

**Figure 5.15**   *Apparent peak loading*

In the figure, it is seen that some peaks deflect even though they are not contacted. Contact factors other than geometry have to be considered, for example, the elasticity of the material. It is true that the curvature of the peak is important at any specific contact, but so also are chemical films such as oxides or sulphides on the surface, which are vitally important in, for example, electrical conductivity. Neither random process analysis nor space frequency functions help.

It may be that for tribological situations involving loading and movement of surfaces, even the definition of what a peak is may have to be changed. The reason is given below in brief.

Consider Figure 5.16 (a):

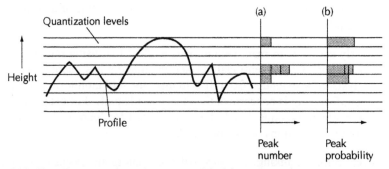

**Figure 5.16**   *Use of histograms for peak counting and dual distance recording*

According to Figure 5.16 (a), the peaks at each level are just counted and then put in the count histogram. This is conventional for communication applications. Notice that no effort is made to find out how long the peak is in the discriminating zones. In Figure 5.16 (b), the dwell distance is measured rather than the count. The histogram of this in (b) is quite different from that in (a).

Hence, calculation of mean peak height and peak moments are quite different using these two methods. For example, the mean height using count only as in (a) is $\dfrac{\pi}{-2} R_a$ higher than the mean height in (b). This represents a completely different situation especially, for example, in load-carrying capacity: communication theory is simply not adequate for tribology!

There is growing evidence that when using this new philosophy, the material ratio concept should be replaced by one in which the new peak definition is used.

i.e. $MR^{\bullet}(\%) = MR(\%) \times$ peak factor

Figure 5.17 shows the different curves for the different methods.

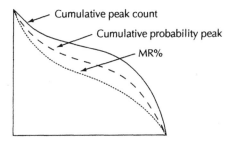

Cumulative peak count

Cumulative probability peak

MR%

**Figure 5.17**    *Alternatives to material ratio*

The problem with peak definition is symptomatic of the way in which contact and tribology function have been neglected in the subject development. The comments above are the top-down-bottom-up requirement to make tribology prediction credible.

It is not catered for in communication theory. In addition to the top-down problem there is the serial-parallel problem. This is concerned with the fact that in contact theory, the contacts can be anywhere in an area – at any time. It is a parallel processing problem. This also is not catered for in communication theory. There is no equivalent in communication theory so it is not surprising that this tribological problem has not been covered.

There is one redeeming property of the random process approach, at least for small movements. This unexpected bonus concerns the properties of the gap between two surfaces. It turns out that the correlation function and the power spectrum have superposition properties if the variables are small. An example of this is given in Figure 5.18. This uses the fact that correlation and spectral analysis are linear processes and are therefore additive.

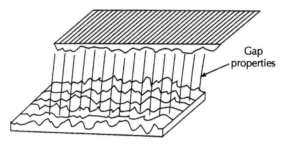

**Figure 5.18**   *Properties of gap between two surfaces*

$$C(\tau) = C(\tau) + C(\tau)$$

$$P(w) = P_1(w) + P_2(w)$$

The autocorrelation and the power spectrum are additive.

## 5.4 Summary of function

**Table 5.2**   *The influence of surface topography on function (from Brian Griffiths) [5.18]*

| Function | Heights | Distribution and shape | Slopes and curvature | Lengths and peak spacing | Lay and lead |
|---|---|---|---|---|---|
| Typical parameters | $R_a\ R_q\ R_t$ | $R_{sk}\ R_{ku}$ | $R_{\Delta q}$ | $R_{sm}$ HSC | Std Sal |
| Bearings | √√ | √√ | √ | √ | √√ |
| Seals | √√ | √√ | √√ | √ | √√ |
| Friction | √√ | √√ | √√ | √√ | √√ |
| Joint stiffness | √√ | √√ | √ | √ | √ |
| Slideways | √√ | √√ | √ | √√ | √√ |
| Contacts (elec/therm) | √√ | √√ | √√ | √√ | |
| Wear | √√ | √√ | √√ | √√ | √√ |
| Galling | √√ | √ | √√ | x | |
| Adhesion & bonding | √√ | √√ | √ | √ | √ |
| Plating & painting | √√ | √ | √ | √ | |
| Forming & drawing | √√ | √ | √ | √√ | √ |
| Fatigue | √√ | √ | x | x | √√ |
| Stress & fracture | √√ | x | √ | | √√ |
| Reflectivity | √√ | | √√ | √√ | √√ |
| Hygiene | √√ | √ | √ | | |

Key: √√ – Much evidence √ – Some evidence x – Little or circumstantial evidence

**Table 5.3**   *Relation between Motif parameters and function of surfaces [5.19]*

| Surface | | Functions applied to the surface | | Roughness profile | | | Waviness profile | | | | Primary profile | |
|---|---|---|---|---|---|---|---|---|---|---|---|---|
| | | Designations | Symbol*) | R | Rx | AR | W | Wx | Wle | AW | Pt | P&c |
| two parts in contact | with relative displacement | Slipping (lubricated) | FG | ● | | | ≤ 0,8R | | | O | | ● |
| | | Dry friction | FS | ● | | O | ● | | | O | | |
| | | Rolling | FR | ● | | | ≤ 0,3R | ● | | O | | O |
| | | Resistance to hammering | RM | O | | O | O | | | O | | ● |
| | | Fluid friction | FF | ● | | O | | | | O | | |
| | | Dynamic sealing — with gasket | ED | ● | O | O | ≤ 0,6R | ● | | O | | |
| | | Dynamic sealing — without gasket | | O | ● | | ≤ 0,6R | | | | | ● |
| | without displacement | Static sealing — with gasket | ES | O | ● | | ≤ R | | O | O | | |
| | | Static sealing — without gasket | | O | ● | | ≤ R | | ● | | | |
| | | Adjustment without displacement with stress | AC | O | | | | | | | | ● |
| | | Adherence (bonding) | AD | ● | | | | | | | O | |
| Independent surface | with stress | Tools (cutting surface) | OC | O | | O | ● | | | ● | | |
| | | Fatigue strengths | EA | O | ● | O | | | | | | O |
| | without stress | Corrosion resistance | RC | ● | ● | | | | | | | |
| | | Paint coating | RE | | | O | | | | O | | |
| | | Electrolytic coating | DE | ● | ≤ 2R | ● | | | | | | |
| | | Measures | ME | ● | | | ≤ R | | | | | |
| | | Appearance (aspect) | AS | ● | | O | O | | | O | | |

Most important parameters: specify at least one of them.
Secondary parameters: to be specified if necessary according to the part functions.
The indication ≤ 0,8R, for example, means that, if the symbol FG is indicated on the drawing, and W not otherwise specified, the upper tolerance on W is equal to the upper tolerance on R multiplied by 0,8.

\*) The symbols (FG, etc.) are acronyms of French designations.

Summarizing, there is now a possibility of weaving a coherent thread linking manufacture metrology and function. This thread *starts* with the function map whose format is outlined above with gap and relative velocity axes. Superimposed on this map is the surface metrology template (or dimensional template) whose axes are *static parameters* e.g. peaks usually from a profile, against *dynamic parameters*, which involve slopes and curvatures and have to be areal. The manufacturing template has 'process' and 'machine tool' as axes. This is superimposed onto the map and metrology template.

Ideally, reading through the template stack (Figure 5.19) should give the best surface and manufacturing process for a particular function. This is shown pictorially above but would be carried out from stored computer data bases.

A further step, which is now a possibility, is to forget the parameters of the surface and carry out a 'pilot' experiment in the computer to see if the workpieces work together. This involves areal mapping of both surfaces comprehensively using a tactile sensor and then literally making them contact and rub by simulation. It may be that this is the best way forward. Notice that the tactile sensor is suggested as the

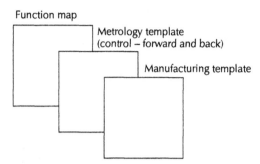

**Figure 5.19**    *Arrangement of templates – the template stack*

preferred instrument. The reason for this is that, in the tradition of metrology, the measurement should mimic as nearly as possible the function. Because most applications involve contact and rubbing, the tactile stylus method should be used. For non-contact applications, obviously optical or other methods are to be preferred.

Chapter 5 reveals an unsatisfactory state of affairs. The actual amount of information linking surface texture to performance is minimal. Only 'types' of parameter e.g. extremes or averages can be safely associated with different types of function. Tables such as 5.2 and 5.3 do not help much. The fact that there are so many ticks indicates that only types of parameters are meaningful. That this important point is not stated suggests that a lot more work needs to be done. It is hoped that this chapter is a step in this direction.

It is postulated here that one of the biggest problems has been the weakness in specifying the objective properly. The 'function' part of the problem has been almost completely neglected. This is why an attempt has been made here to characterize function by means of a map. The spatial, temporal and force regimes are absolutely essential in order to be able to understand what is happening. This is the reason why Figures 5.1 and 5.3 have been introduced in this section. When the desired characteristics of the function have been correctly identified, then and only then will it be possible to make use of the investment made in parameters by the instrument companies. Still to be achieved is the characterization of the system i.e. the two surfaces.

## References

5.1   Whitehouse, D. J. *Handbook of Surface Metrology.* Institute of Physics Publishing Bristol and Philadelphia (1994).

5.2   Schlesinger, G. *Surface Finish.* Institution of Production Engineers London (1942).

5.3   Whitehouse, D. J. *Surfaces, a Link between Manufacture and Function.* Proceedings Institute of Mechanical Engineers 192 179–188 (1978).

5.4   Wigner, E. *On the Quantum Correction for Thermodynamic Equilibrium*. Physics Review 140 749–759 (1932).

5.5   Woodward, P. M. *Probability and Information Theory with Application to Radar*. Pergamon, New York (1953).

5.6   Zheng, K and Whitehouse, D. J. The *Application of the Wigner Distribution to Machine Tool Monitoring*. Proceedings Institute of Mechanical Engineers 206 249–264 (1992).

5.7   Daubechies I. *The Wavelet Time-frequency Localisation and Signal Analysis*. IEEE Transactions on Information Theory 36 961–1005 (1990).

5.8   Strang, B. and Nguyen, T. *Wavelet and Filter Bands*. Wellesley Cambridge Press (1996).

5.9   Perthen, J. *Prufen und Messen der Oberflachengestalt*. Carl Hanser Munich. (1947).

5.10 Greenwood J. A. and Williamson J. B. P. *The Contact of Nominally Flat Surfaces*. Burndy Research Report 15 (1964).

5.11 Archard, J. F. *Elastic Deformation and the Laws of Friction*. Proceedings Royal Society London A 243 190–205 (1957).

5.12 Whitehouse, D. J. and Archard, J. F. *Discrete Properties of Random Surfaces of Significance in their Contact*. Proceedings Royal Society London A 316 97 (1990).

5.13 Whitehouse D. J. *Some theoretical Aspects of Surface Peak Parameters*. Precision Eng. 23 94–102 (1999).

5.14 Stout, K. G. *Development of a Basis for 3D Surface Roughness Standards*. EC Contract No. 5MT4-CT98-2256.

5.15 Euro Report 15178EN.

5.16 Scott, P. *Areal Topological Characterization for a Functional Description of Surfaces*. European Surfaces Workshop Corps La Salette June (1998).

5.17 Radhakrishnan V. *On an Appropriate Radius for the Enveloping Circle for Roughness Measurement in the E system*. Annals CIRP 20 No.1 (1971).

5.18 Griffiths B. Private Communication (1999).

5.19 ISO Handbook 33 (1998).

5.20 Whitehouse D. J. *Dynamic and Systems Parameters for Surface Evaluation* (in press).

# 6

# Surface finish measurement – general

## 6.1 Some quick ways of examining the surface

### 6.1.1 The surface itself

Some parameters can be estimated without using an instrument. In fact, before instruments were generally available, the surface was examined quite successfully yet subjectively with the eye and the fingernail. Never denigrate this approach. Far too many people, including designers, do not look at the surface enough.

Figure 6.1 (a) shows the light scatter from a surface when a ray of light is thrown onto it. The light is scattered over a range of angles. However, if the incident ray and the eye are positioned at a glancing angle off the surface (Figure 6.1 (b)) the surface can behave as if it is a mirror. This is called Lambert's Law. Why it does this is indicated in Figure 6.1(c).

### Roughness evaluation: Lambert's law

From Figure 6.1 (c), the path difference is 2x between rays off peaks (path 1) and between rays off valleys (path 2). From Figure 6.1 (c), $R_t \sim \dfrac{\lambda}{8 \sin \theta_L}$.

The surface is tilted until it acts like a mirror. This occurs at a small glancing angle $\theta_L$. The difference in path length of rays hitting peaks (as path 2) in the figure and rays hitting valleys (as path 1) is approximately $\lambda$, where $\lambda$ is the wavelength of light ($\sim 0.6\mu m$).

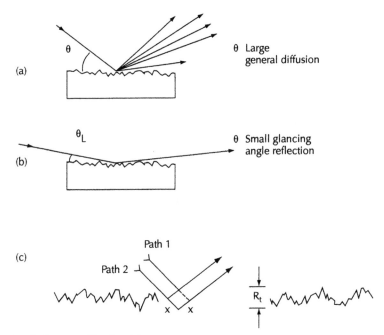

**Figure 6.1**   *Applications of Lambert's law*

For example, if $R_t$ is 1µm, the Lambert angle $\theta_L$ is $\sin^{-1}\left(\dfrac{\lambda}{8R_1}\right) \sim 4^0$. Estimating this

angle by eye when handling the workpiece is not very accurate but it does enable an estimate of $R_t$ to be made.

## Directionality

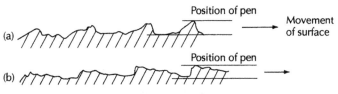

**Figure 6.2**   *Movement across surface to detect surface quality*

The surface shown in Figure 6.2 (a) can have the same roughness as the surface in Figure 6.2 (b) but the two surfaces behave quite differently in a functional situation, say, moving left to right in a bearing.

An experienced inspector or assembler can detect which way the 'rasp-like' pattern lies with respect to the proposed movement by using the fingernail. The apparent friction of (a) is greater than (b). An even simpler and yet less subjective approach is to run a ball point pen across the surface (as if it is being written on). Situation (a) 'squeaks' louder than (b). In the figure, the surface is moved.

## Lay pattern (isotropy)

A great deal of theoretical work has and is being carried out to find a way of measuring and characterizing the isotropy (the lay) of the surface (see Chapter 2). Note that the lay is determined from the path of the tool or process movement. It is obvious from the machine tool process what the lay is but not necessarily from a post-process inspection, where the tool movement is possibly unknown. The lay can quite easily be detected. Quantifying it is another matter. It is the value and character of lay that is important in functional prediction.

Figure 6.3 shows a machined surface which has been cleaned, usually by acetone after which a drop of oil is put onto the surface. This obviously produces a 'blob' on the surface as shown, which spreads according to the lay.

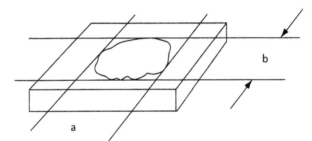

**Figure 6.3** *Application of oil drop to surface*

The dimension of 'a' relative to 'b' is an indicator of isotropy and can be used as a measure. For an isotropic surface a = b, whereas for an anisotropic surface a > b or b < a.

In the direction 'a', the smallest diameter, the surface marks are close together; in 'b' they are further apart. In the 'a' direction, the surface slopes are high because the roughness is 'bunched'. In 'b' they are stretched.

The principle behind this is based on surface tension. Figure 6.4 shows a bubble on a surface.

**Figure 6.4** *Angle of meniscus*

If is does not 'wet' the surface, a fixed angle α is produced between the machined surface and the meniscus; wetting means spreading indefinitely.

Because the slopes are bigger in the direction 'a', the criterion for α is easily obtained but for 'b' the film has to spread more widely to satisfy the angle α. This criterion has to be achieved at the rim of the meniscus, which becomes in effect a contour of fixed angle on the surface.

Incidentally, the 'a' and 'b' directions can also be found using a 'biro'. In 'a' the squeak is loudest, and in 'b' quietest. This method is easy to carry out but does not produce a measurable record, whereas the oil method does (Figure 6.4).

If the meniscus edge is indistinct, it can be highlighted by inserting a light source in the drop – for example by means of a fibre optic light probe. It does not matter where the probe is placed. The edge is revealed as shown in Figure 6.5 (a) and (b).

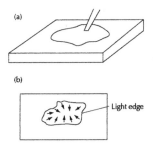

**Figure 6.5 (a)**   *Oil drop spread*

### 6.1.2   The profile record

It is possible to use the acuity of the eye to gather information from the chart. To take advantage of the properties of a visual examination, the chart length should be no longer than a metre. Any chart length longer than this means that the eye cannot absorb all the data at a glance.

It could be argued that the chart is redundant, given the comprehensive list of parameters available from instruments. This is not so. The profile contains enormous amounts of data. Obviously, the actual roughness value is determined by calibration of the instrument but other parameters such as the statistical parameters skew and kurtosis can be examined from the chart. For example, the presence of skew can be revealed simply by turning the chart over (Figure 6.5 (b) overleaf).

Only skew is instantly revealed by this. Also, this operation serves as a check on the actual value of the skew. This is possible because a freak peak or valley can completely dominate the numerical skew value yet it is not typical of the

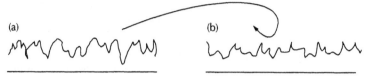

**Figure 6.5 (b)**   *Visual skew*

machining process. The instrument computer can be fooled by these freak peaks, but the eye instantly filters them out.

Kurtosis can also be checked for simply by foreshortening the chart by viewing it at a glancing angle as shown in Figure 6.6 (b) rather than normally as shown in Figure 6.6 (a).

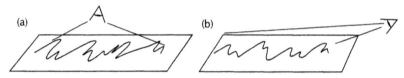

**Figure 6.6**   *Visual kurtosis*

Any 'spikiness' is immediately amplified by this method; 'bumpiness' is not shown up to the same extent so that the presence of high kurtosis – which is usually detrimental to performance – can be spotted.

As in skewness, freak behaviour will be disregarded by the eye.

Another point concerning comparisons or estimates of parameters is that the same part of the surface should always be used. If two profile charts from different parts of the surface are compared, any differences are probably due to statistical variation and are not necessarily significant. Exact relocation of the profile has to be made, for example, if wear is being monitored. If the surface being worn is always measured in the same place, it is possible to quantify the wear point by point on the profile simply by overlaying the worn profile onto the original as seen in Figure 6.7.

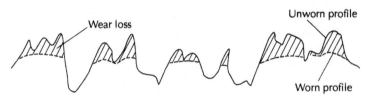

**Figure 6.7**   *Use of relocation to measure wear*

The wear in this position is the shaded area on the overlaid chart (Figure 6.7).

## 6.2   Surface finish instrumentation

### 6.2.1   Issues

There are a number of issues for measuring texture as well as roundness. Some things are common. The first issue is to decide what property is required and, more to the point, what is being measured. The following are points to be considered.

1.   (a)   Stylus methods – elastic and plastic as well as geometrical properties are involved.

  (b)   Optical methods – optical path length measured, not geometry.

  (c)   Scanning microscopes – very wide range of unknown variables.

2.   Can the instruments be calibrated and can this 'calibration' be traced back to national and international standards?

3.   Is the instrument versatile and robust enough for general engineering use or is it more suitable for research laboratories?

4.   If for engineering use, where will it be placed and when will the measurement be taken?

5.   What form will the output take, and what processing will be necessary?

### 6.2.2   Nature of measurement

The possibilities are listed below in Table 6.1.

**Table 6.1**   *Types of measurement and their properties*

| Where? | In-process | In situ | Remote |
|---|---|---|---|
| What type of measurement | Dynamic | Static | Integrated |
| Type of control | Adaptive control | Statistical control | Long-term traceability |
| Information obtained | Very specific | Medium | Comprehensive surface geometry |
| Information required | Process control | Process monitor | Machine tool monitor |
| Speed | Very fast<br>No operator intervention | Time to record and judge | Functional judgement |
| Outcome | Working shift controlled | Quality of conformance assured | Quality of design assured |

Production Assurance                    Total Quality Assurance

### 6.2.3 Historical

It has been known for some time that surfaces are important in the manufacture and performance of workpieces. The problem is that there are a multitude of requirements that are not necessarily compatible with each other.

The investigation of surfaces is an old technology. Leonardo da Vinci, Amonton and Coulomb were all interested in surface behaviour. However, their investigations were largely subjective and not much use.

In the twentieth century, perhaps the first attempt to make an instrument was Tomlinson's in 1919 [6.1] at the NPL. He devised a crude lever system, which magnified the surface profile mechanically by a factor of about 20:1. This gave him just enough magnification to be able to see the surface detail but not big enough to work on.

Surface instrumentation was really started by Schmaltz who in 1929 developed the first stylus instrument and began experimenting with optical methods. His philosophy was very simple. He had seen men in the production departments of some large firms e.g. General Motors and Ford scratching surfaces with the fingernail to assess the roughness and sometimes viewing the surface obliquely under strip lighting. Schmaltz's idea was to replace use of the fingernail with the stylus and the eye with imaging optics [6.2].

Neither instrument was practical but nevertheless they stimulated Dr. Abbott of Physics Research Inc. in 1936 [6.3] to advance the technology by making the output from the probe electrical so that substantial magnifications could be achieved. This output was put on a meter for all to see. It still lacked a permanent record for use on the shop floor. It was R.E. Reason of Taylor Hobson in 1939 [6.7] who completed the jigsaw by adding on a chart recorder. Hence, the birth of the profile graph, which was enabled, by having a tracking mechanism to move the stylus across the surface.

Optical methods did not, as a rule, follow this pattern of development [6.4]. The preferred way was to use fringe methods notably by Linnick [6.11] (1944) and Tolansky in1946 [6.12]. In this way, all of an area of surface could be examined simultaneously, which was an advantage but the output could only, in those days, be viewed.

So, stylus and optical methods developed along different paths; the stylus method being used for height information and the optical methods for areal information.

Providing that the applications were different, these two separate paths were acceptable. Problems began to arise when both methods were used on the same object. Users expected the readings to agree: a requirement sometimes not possible.

Alternative methods such as capacitance [6.5] and pneumatic devices [6.6, 6.8] have also been developed at the same time but with less success.

## 6.3   Comments

### 6.3.1   Stylus methods

The basic features of the stylus instrument are:

(a)  the probe;
(b)  the transducer and reference;
(c)  the amplifier;
(d)  filter or processor;
(e)  recorder or other output such as a meter or computer.

### 6.3.2   The optical method

A whole range of optical methods will be examined to see which of them can fit different applications. It will be shown that the optical methods are versatile but suffer from noise arising from their physics rather than by any instrumental reasons.

In order to clarify the choice of technique, comparisons will be made showing the advantages and disadvantages of stylus and optical methods and their signal to noise ratios.

### 6.3.3   Other instruments

Other instruments will be discussed briefly later on. These include the new generation of scanning microscopes such as the **Scanning Electron Microscope (SEM)**, the **Scanning Tunnelling Microscope (STM)** and the **Atomic Force Microscope (AFM)**.

### 6.3.4  Instrument comparisons

There are two methods of comparing instruments for surface texture (see Figure 6.8). These are not necessarily the same for roundness/cylindricity.

There is an unsatisfactory situation regarding instrument criteria. This is because the criterion that is most often referred to concerns instrumental specifications, which are most useful in research [6.1]. An example is shown in Figure 6.8 (a). The axes here are surface amplitude measurement capability as one axis and surface wavelength measurement capability as the other. It is easily possible to plot instrument specification on this basis. However, research use is hardly the most useful criterion for the majority of users. Figure 6.8 (b) shows an alternative plot, which has frequency response or equivalent as one axis and the other range/resolution [6.2]. This is aimed at industrial use. Effectively, the ordinate axis shows the suitability of the instrument for fast measurement e.g. in-process gauging, whereas the abscissa shows the extent to which the instrument can be used in integrated measurement. Unfortunately, this latter plot is rarely seen because day to day industrial use rarely makes the headlines!

How instruments compare can be determined from the two plots in Figure 6.8.

It is necessary to differentiate between the types of instrument, because in the industrial case the time to get an output can be important but it is not so important in research. What is important is to be able to measure the surface and to be sure that (a) it does not damage the surface and (b) it has high fidelity. Both of these issues will be considered in Section 7 and other instrument sections.

For interest, Plate I shows some early instruments; (a) shows Abbott's Profilometer, (b) the Schmalz photomicroscope and (c) Taylor Hobson surface meter (before the Talysurf name).

## References

6.1  Tomlinson (1919) – see, for example, Schlesinger G. Machinery. Vol 55 p721 (1940).

6.2  Schmalz G. Z. VDI. Vol. 73 p144–161 (1929).

6.3  Abbott J. and Firestone A. *A New Profilograph Measures Roughness*. Autom. Ind. P204 (1933).

6.4  Reason R. E. *Biographical Memoirs by D. J. Whitehouse*. Me. Proc. Roy. Soc. Vol. 36 p437 (1990).

6.5  Linnich W. *Ein Apparat zur Messung von ver Schiebungen in der Schrichtung*. Z. Instrum. Vol. 50 p192 (1930).

6.6  Tolansky S. *Vertrag*. Tag. Deutsch Phys. Ges. 5–7 Gottenge (1941).

(a) Research criterion

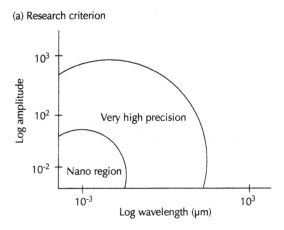

(b) Industrial criterion

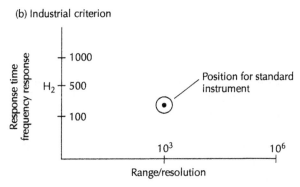

**Figure 6.8**   *Instrument criteria*

6.7   Sherwood K. F. and Crookall J. R. *Surface Finish Measurement*. Proc. Inst. Mech. Eng. Vol. 182 pt 3k (1967/68).

6.8   Greneck M. and Wunsch H. L. *Application of Pneumatic Gauging*. Machining. Vol. 81 p707 (1952).

6.9   Stedman M. *Mapping the Performance of Surface Measuring Instruments*. Proc. SPIE. Vol. 803 p138 (1987).

6.10 Whitehouse D. J. *Handbook of Surface Metrology*. Inst. Phys. Pub., Bristol (1994).

# 7

# Stylus instruments

## 7.1 The stylus

The system has been featured in Section 6.3.1, the central component being the stylus. This was originally the fingernail until Schmaltz [7.1] pushed the idea forward to a mechanical probe [7.2]. The different types are shown in Figure 7.1.

In the UK, there was first a diamond stylus of 90° pyramid shape. This had two dimensions at the tip; one of 2μm and the other of about 7μm. The 2μm edge traversed across the lay. The 7μm dimension was intended to give the stylus tip some mechanical strength . This worked well on surfaces with an appreciable lay but not at all on isotropic surfaces. The 7μm dimension dictates the overall resolution limit.

In the US a diamond cone was used, which had an angle of 60°. The result was that the two countries developed instruments that had different starting points, which caused some trouble with cross-calibration exercises.

It has been pointed out that the two diamond shapes are not crucial because, in practice, the cone gets a flat on it with use and the pyramid gets rounded corners with use. Although this is true for the tip shape, there is still a different angle that could make a difference on fine surfaces.

Figure 7.2 shows that the stylus slope has to be considered. The tip can be infinitely sharp but still will not penetrate completely into the groove or valley. The slope effect occurs more for fine surfaces where the $R_a$ is less than 0.1μm.

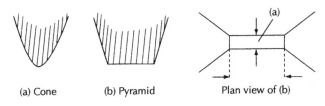

(a) Cone        (b) Pyramid        Plan view of (b)

**Figure 7.1**  *Stylus shape*

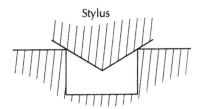

**Figure 7.2**  *Stylus slope effect*

## 7.2  Reference

It has been mentioned in Chapter 2 with regard to characterization that all instruments are, in fact, callipers. One arm makes contact with a reference and one with the surface under test. Figure 7.3 shows that establishing the reference is quite difficult in surface metrology. It is usually a two stage process. The first is to adjust a mechanical reference that is smooth and straight and upon which the reference arm rests. The adjustment is carried out until the signal is within the range of the transducer over all its traverse. The next step is to derive a reference from the collected data. This computed reference has to be placed within the profile signal (Figure 7.4). It can take care of tilts or curves but it has to be remembered that it is computed. It is in effect a virtual reference and that serious errors can occur in its generation.

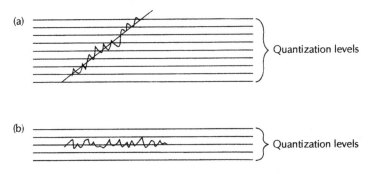

**Figure 7.3**  *Software reference*

In Figure 7.3, it can be seen that if the slope is extreme the magnification has to be low to keep the trace on the chart. When the tilt is removed computationally, the resultant roughness signal only occupies a few quantization levels with the result that the roughness is likely to be contaminated with numerical noise (Figure 7.4). Figure 7.5 shows how a curved reference can be computed that presents exactly the same problems.

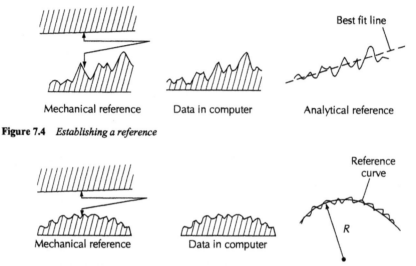

**Figure 7.4**    *Establishing a reference*

**Figure 7.5**    *Fitting a curved reference – actual case 1*

## 7.3   Use of skids

Because of the time taken to adjust the mechanical reference into place, short cuts have been devised [7.4]. The most popular is what is called an **intrinsic reference**. It replaces the flat smooth remote mechanical reference. Instead, the reference calliper arm sits on the same specimen as the test calliper stylus of the reference calliper arm called the **skid**. The only difference is that the skid is much blunter than the sharp stylus so that it tends to bridge the gap between peaks on the surface. It can never generate a straight reference like the mechanical one but, providing that there are constraints on the radius and the spacing between the stylus and skid, it is acceptable. Incidentally, the skid principle is also used in optical instruments.

Figure 7.6 shows how the skid can distort a surface signal. Figure 7.7 shows problems that can be caused when the skid-stylus separation is a fraction of a wavelength on the surface. The effective bandwidth range where it is acceptable to use a skid is shown in Figure 7.8. As can be seen, it is quite a small band relative to the wavelength range of the surface.

Another problem with using a skid is that it can in certain circumstances damage the surface, especially if debris gets trapped between the skid and the surface. It has to be remembered that the skid supports the whole weight of the traverse unit. The stylus itself is very lightly loaded.

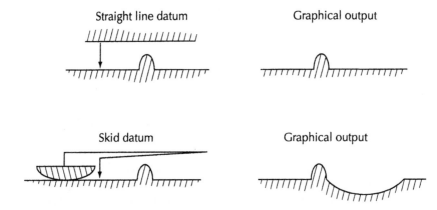

**Figure 7.6** *Effect of skid on surface signal*

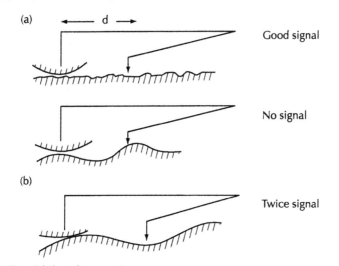

**Figure 7.7** *Effect of skid – stylus separation*

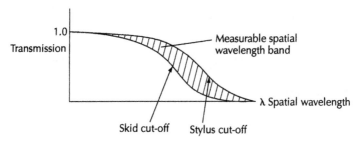

**Figure 7.8** *The measurable surface bandwidth [7.5]*

## 7.4　Pick-up systems

Two types of pick-up are in common use. That shown in Figure 7.9 is called a **side acting gauge**. It is this configuration because very many engineering parts have holes in them. The problem with this is that the resonant frequency of the pick-up can be quite low e.g. 200Hz, which can restrict the use of such instruments if speed is required. It is possible to improve the frequency effectiveness but these methods are not in general use.

Figure 7.10 shows an alternative pick-up that has a considerably higher frequency response but which is somewhat restricted as a side acting gauge. Also, the force presented to the surface changes with deflection because it is a short cantilever. Frequency responses of over 1KHz can be achieved using this configuration.

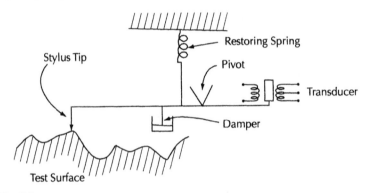

**Figure 7.9**　*Side acting gauge*

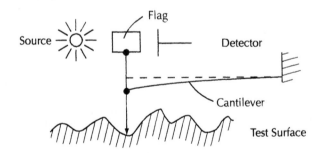

**Figure 7.10**　*Cantilever mode*

## 7.5　Stylus damage

Sometimes the stylus method is criticized because it is claimed to damage the surface. In most cases if this is reported, it is usually the skid that is the culprit and not the stylus. With soft surfaces, it is never advised to use the skid method. It is, however, quite easy to derive a criterion for the stylus so that the surface is not damaged. The steps taken to obtain a suitable criterion are shown in Table 7.1.

It emerges that the damage index $\psi$ so defined has to lie between acceptable limits if no damage is to be done. This criterion is [7.2].

$$\psi = \frac{1}{\pi}\left(\frac{W}{H^3}\right)^{\frac{1}{3}} \cdot \left(\frac{E}{R}\right)^{\frac{2}{3}} \frac{1}{(1.1)^2} \tag{7.1}$$

In this formula $W$ is the load, $H$ is the hardness of the material being measured, $R$ is the radius of the tip of the stylus, $E$ is the elastic modulus.

For $\psi > 1$ there can be damage because the elastic pressure is greater than the hardness. For $\psi < 1$ there is no damage.

In fact, the difficult parameters above are $W$ and $H$. $W$ is the effective mass of the pick-up as it traverses the surface. The problem is that it should include the dynamic forces imposed on the stylus by the surface and consequently reacts with the surface by the same amount. The maximum value of the dynamic force is $2W$ where $W$ is the rest weight. This force acts in the valleys so damage, if any, is due to the stylus and will be a maximum in the valleys – not the peaks. Peak damage is due to skid.

The other difficult parameter is the value of $H$. Unfortunately, most people think that the bulk value of $H$ should be used in calculations like this i.e. that value of hardness obtained when the surface is indented by tens of micrometres. This value of hardness is incorrect. The value of hardness should be the skin hardness. This is the value obtained when the indentation measures fractions of a micrometre. In such cases, the effective hardness is two or three times the bulk value. For copper, for example, it could easily be 300 VPN rather than the bulk value of 100 VPN.

Taking these factors into account and bearing in mind the sometimes improper use of a skid, it is found that the stylus rarely causes damage. If there is any doubt, then one preventative measure that can always be made is to lower the dynamic force by reducing the speed of traverse of the system. Note that the skid damage is on peaks, whereas the stylus damage is in the valleys.

## 7.6   Stylus instrument usage

### 7.6.1   Measurement issues

One rule in metrology is that the number of instrument types in any application should be kept to a minimum. Failure to do this results in errors of cross-calibration e.g. getting optical readings and stylus readings to agree on the same surface.

**Table 7.1**  *Surface damage index*

| |
|---|
| Using the Hertz approach<br><br>$\text{Elastic area} = \pi a^2 = \left(\dfrac{4WR'}{3E}\right)^{\frac{2}{3}}$<br><br>where $W$ = load, $E$ = elastic modulus, $R$ = radius<br><br>$\text{Pressure} = \dfrac{W}{\pi a^2}$<br><br>If $H$ is the hardness and $\psi$ the index<br><br>$\psi H = \dfrac{1}{\pi}\left(\dfrac{W}{H^3}\right)^{\frac{1}{3}}\cdot\left(\dfrac{E}{R}\right)^{\frac{2}{3}}$<br><br>So<br><br>$\psi = \dfrac{1}{\pi}\left(\dfrac{W}{H^3}\right)^{\frac{1}{3}}\cdot\left(\dfrac{E}{R}\right)^{\frac{2}{3}}$<br><br>If $\psi > 1$ elastic pressure $P_e > H$<br>The surface is scratched<br>$\psi < 1$ no damage<br>the question is what value of $H$ has to be used<br><br>A practical example of the stylus damage index is given below.<br><br>The index $\psi$<br><br>$\psi = \dfrac{W^{\frac{1}{3}}E^{\frac{2}{3}}}{\pi \times (1.1)^2 \times R^{\frac{2}{3}} \times k \times H}$<br><br>where $k$ is the ratio of microhardness to bulk hardness for the material. Here it is taken as 3. The material is copper.<br><br>Hence typically $\quad W = 50\mu N = 50\times10^{-6}N$ (Newton)<br>$\qquad\qquad\qquad\ E = 130\text{GPa } [1\text{giga Pascal} = 10^9 \text{ N/m}^2]$<br>$\qquad\qquad\qquad\quad = 130 \times 10^9 \text{ N/m}^2 \text{ [Newton/metre squared]}$<br>$\qquad\qquad\qquad\ H = 150 \text{ DPN (Vickers hardness in kg/mm2)}$<br>$\qquad\qquad\qquad\quad = 1.5 \times 10^9 \text{ N/m}^2$<br>$\qquad\qquad\qquad\ R = 10\mu m = 10^{-5}m$<br><br>Hence $\psi = 0.59$ for this set of values and the stylus should therefore not damage the surface. |

In engineering, one of the problems is that environmental conditions can vary widely between the machine shop and the inspection room. This is where the stylus method scores over other methods. The fact that the stylus touches the surface can be an advantage in two ways. The first is that the stylus can push aside debris and penetrate films that would otherwise mask the geometry. The second is that the stylus can be used to act upon the surface, as in the case of the scanning microscopes, to initiate discharge, disrupt fields and behave in a proactive way as required. It becomes part of the experiment rather than an observer. This aspect of stylus (or probe) usage has extended the application of the stylus method to encompass other fields. Also, in general engineering applications, there is evidence [7.6] that tribological aspects of the surface such as friction, nano-hardness and nano-elasticity can be examined.

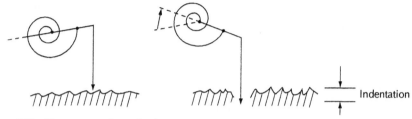

**Figure 7.11**    *Measurement of nano-hardness*

Figure 7.11 shows how the nano-hardness and nano-elasticity can be measured. The instrument to use is the Talystep or Nanosurf or similar instrument that has an adjustable load on the stylus. The probe is positioned on the surface as if a roughness trace is to be made i.e. the recorder pen is near the centre of the chart. Then a known force $W$ is applied by means of the adjustable load control. The deflection of the probe into the surface can be measured by the pick-up transducer. Let this be $D_1$. Then the load is released and the recovered pen movement measured. This is $D_2$. The difference between $D_1$ and $D_2$ is the nano-hardness (measured on the Rockwell scale). The value $D_2$ is the nano-elasticity.

Two things need to be known. One is the geometry of the stylus and the other is the elasticity in the metrology loop of the instrument. The latter is found by setting the stylus down onto a very hard material such as topaz when $W$ is applied as before. Any deflection measured by the transducer is the latent elasticity of the instrument, which should be taken away from the readings.

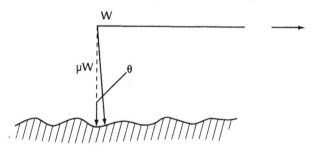

**Figure 7.12**    *Measurement of friction*

The deflection of the probe shank in the lateral direction can be measured and used to estimate $\mu$, given the stylus probe force $W$ as shown in Figure 7.12. Obviously, the material of the probe has to be selected carefully to fit the application [7.7].

The ability of the contacting stylus to measure geometry in the presence of debris and other extraneous elements makes it ideally suited to use when the application of the surface involves contact. Most automotive applications fall into this category; gears, cams, pistons etc. are but a few. This is why the stylus method has been widely adopted. Because of the wide variety of shapes and sizes of parts, a family of instruments has been developed.

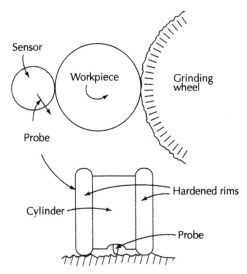

**Figure 7.13**  *Deutschke method for in-process measurement*

### 7.6.2  In-process measurement

The only serious attempt to use a stylus technique in an in-process mode was due to Deutschke in Stuttgart [7.8]. The instrument was not meant to give a continuous output but more of a sampled output.

Figure 7.13 shows the set-up. The basic part of the instrument consisted of a drum with a hole in the side through which poked a transducer. This was held in radially by means of a spring. The drum contacted the workpiece. Every revolution of the workpiece turned the drum. Centrifugal force pushed the transducer out of the drum onto the workpiece. At this stage, the radial position of the transducer was estimated by an electrical bridge system within the drum. Hence, as the work was being machined, the transducer produced a set of point heights of the workpiece finish, which could be analysed to give a height-frequency of occurrence distribution from which $R_a$ and $R_q$ could be gauged. The system was designed to measure ground surfaces. It was never successful because it tended to indent the surface in a rather unsatisfactory way rather than to just make contact with the surface.

It is not surprising that the method failed. In-process measurement invites a continuous mechanism. The discrete point by point method requires stable conditions during the measurement cycle i.e. the workpiece is stopped.

To have the stylus resting on the surface to provide a continuous signal is unrealistic because the radial dynamic force on the stylus can throw it off the surface. Once the dynamic forces equal the static forces, not only can lift-off occur at the peaks but twice the static load is applied to the surface in the valleys. Both of these effects impair fidelity and have to be avoided. Unfortunately, the dynamic forces depend on rotation of the workpiece and are not under the control of the instrument maker, which can cause operational problems with filters and sampling rates.

In-process measurement using stylus methods is difficult, if not impossible. Optical methods are also difficult to apply – not because of the forces but due to the presence of coolant and debris that the optical method tries to measure.

### 7.6.3   In situ measurement – hand-held instrument

The most similar methods to stylus systems are the in situ methods, which take the instrument to the part but measure only when it is stationary.

An in situ instrument usually has to be portable and at the same time be capable of measuring surface parameters that have been shown to be useful in process control.

Plate II (a) shows such an instrument [7.9 (a)], which is small and light and measures $R_a$ and $R_z$, the two best-known control parameters. The advantage of this type of instrument is obvious. It is made so that it can stand alone on a workpiece and measure in situ. Plate II (b) shows the instrument Surtronic Duo (Taylor Hobson) measuring the surface of a cylinder head and a piston (Plate II (c)). Notice that the display can be separated from the measuring unit. Communication between the two units is achieved by an infra-red optical link (Plate II (d)).

A typical specification for such an instrument is 200μm for the range of the gauge.

$R_a$, 0.1μm to 40μm (1μ in to 1600 μ in),
$R_z$, 0.1μm to 199μm (accuracy ± 5% reading), transverse 5mm.

The design of an instrument such as that above is not straightforward. This type of instrument is meant to be taken to the workpiece. This usually implies that the workpiece is large or perhaps inaccessible; it does not usually involve very smooth surfaces. The design has to cater for this by being robust and easy to use. Also, because the surfaces are usually rough, lateral forces on the stylus pick-up are high when taking a measurement.

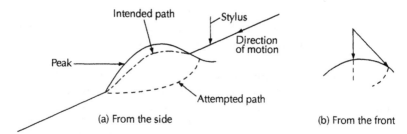

**Figure 7.14**   *Cross-axial problem*

This is due to the high peaks of the rough surface that the stylus has to surmount. The stylus tries to go around large peaks rather than over them as shown in Figure 7.14 (a). The traversing unit supporting the stylus has therefore to have a large cross-axial stiffness. At the same time, stiffness in the measuring direction i.e. normal to the surface has to be as small as possible to be compatible with keeping the stylus on the workpiece during measurement. Stiffness ratios should be at the very least 100:1.

It is difficult to achieve high cross-axial stiffness in a unit that has to be small, light and cheap. Sometimes the simpler instruments are very difficult to design, given the difficult environment they are expected to work in. The very fact that the instrument is portable makes control over its operation difficult.

In order to demonstrate machine capability or to provide records for quality of conformance it is often required to have a record of the results and the profile of the part. See Plate III [7.9 (b)].

A requirement of portable instruments is that they are able to measure in different positions and orientations. For example, it should be possible to measure vertical surfaces and surfaces at odd angles. This requirement automatically rules out styli held onto the surface by gravity. In fact, this type of instrument requires that the restoring spring force be greater than gravity. This extra force adds to the static force when the measured part is horizontal i.e. in the usual measuring position. So, flexibility of operation brings its problems.

Another consideration is the ability to make measurements in holes. Many engineering parts have holes in them. Furthermore, where there is a hole, it is usually functionally important. As a result of these requirements, small stylus instruments often have the side-acting configuration (Figure 7.15).

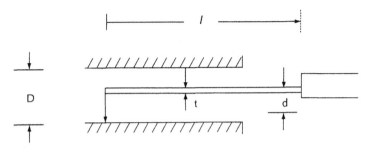

**Figure 7.15**  *Side-acting gauge*

Such systems are penalised by their length *l* and thickness *t*. Long *l*, small *t* and long crank *d* give a low resonant frequency. Various attempts to damp this by making the pick-up hollow and filling it with plastic or similar devices have had only partial success. Optical methods without contact seem attractive but getting the light in and out of the hole is not as easy as with a mechanical movement.

### 7.6.4 Integrated instruments

To offset speed limitations, manufacturers making stylus instruments for general usage have concentrated on measuring as many geometric features as possible with one instrument. So, instruments measuring roughness waviness and form are now available [7.9 (c)]. See Plate III.

The advantage of this integrated approach is that one height calibration should suffice for measuring all the geometric features. This minimizes instrument transfer problems as well as minimizing set-up errors.

It should be remembered, however, that height calibration by itself may be acceptable for long wavelength features on the surface but consideration has to be given to dynamic effects e.g. frequency response when roughness is included. A simple, yet effective, calibration is shown [7.9 (d)] in Plate V(c).

This is a hemisphere of known radius. One track across is sufficient for form waviness and roughness. A prerequisite for such a system is that the transducer has a wide range of height and at the same time has a fine resolution. The first is needed for the form, the latter for the roughness. As a result of these two requirements for the instrument, the division of the two i.e. the range divided by the resolution has become a criterion for integrated instrument performance. This has already been indicated in Chapter 6. A typical range to resolution value is 1000:1. This amplitude criterion is usually augmented by some measure of operating speed. Usually, this is the frequency response of the pick-up or some other limiting response or reaction time. For wide-range stylus pick-ups, 200 to 300 Hz is typical and range to resolution values of $1:10^6$ or thereabouts are now being achieved.

There are two basic problems with this type of instrument. The first is that the pick-up has to swing through a large arc to cope with form or shape variations of 10mm or more. The pick-up wants to measure vertical height rather than angle. This arcuate effect has to be identified i.e. where in the arc the pick-up is and what compensation is needed. The pick-up should have one degree of freedom, ideally translational (Figure 7.16). However, with a side-acting system, the degree of freedom has to be rotational, which can cause small changes in calibration (Figure 7.16 (b)). Some original design is required for this, for example using a phase grating interferometer for the gauge system shown in Plate IV.

The second problem is that the roughness looks very much like noise to the system and so has to be identified and separated from electronic noise e.g. Johnson noise, shot noise and environmental noise such as vibration. Failure to do this can bias the estimate of radius.

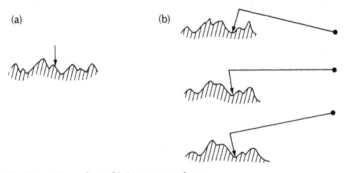

**Figure 7.16**   *(a) Axial transducer (b) Arcuate transducer*

The side-acting system can be extended to measure very large changes of shape and complicated contours. Gauges now exist, which have a range of 50mm in height with resolutions within tens of nanometres. Because of the wide variety of shapes and also the different materials that make up the workpiece, some choice of stylus is made available.

Often, more than one angle of stylus is needed. These could include a 90° chisel or 60° cone or sometimes a sapphire ball; the stylus needs to be matched to the part. A minimum requirement is that the styli have to be easily interchangeable (Plate IV (a)).

Step height measurement for thin films and also measurement of very fine roughness requires extra precautions in operation and design.

The original and unique Talystep design (Figure 7.17) [7.9 (e)] using an arcuate traverse has been replaced by linear traverse units extending the range to 50mm. Environmental chambers are needed to prevent acoustic noise, such as voices, affecting the measurement. Even the small pressure of sound waves can influence the values obtained by such instruments. Failure to isolate sounds can result in the instrument 'talking back' to the operator. This precaution is in addition to conventional anti-vibration measures.

One recent advance has been the use of very stable materials such as Zerodure, which has a low coefficient of expansion. Some consideration also has to be given to reducing the effect of thermal shock by handling or body heat.

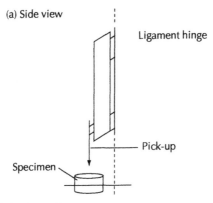

(a) Side view

Ligament hinge

Pick-up

Specimen

**Figure 7.17**    *Original Talystep; traversing*

The arcuate movement was the result of the pick-up being supported on a long ligament hinge. This provided very high cross-axial stiffness and the ability to remove some curvature from the workpiece. Unfortunately, the traverse was rather limited but sub-nanometre resolution was easily achieved. Plate VI shows the instrument now copied by many competitors for measuring microcircuits, photoresists micro-finished metals and ceramics, heat treatment and defect examination. An example is its replacement, the Nanostep, designed by Warwick University and N.P.L. (Plate VII) [7.9 (f)]. Although there are many operational and hardware advances in modern versions of the Talystep, it is in the software where the most benefits have been made. Rarely, however, has there been a more elegant and effective mechanical design as in the original Talystep. So effective was it that the first experiments to measure nano-hardness and nano-elasticity were carried out using it. One reason for this was the variable loading of the stylus (1-70mg) (10-700μN) – a feature that has been preserved [7.6].

### 7.6.5  Fast tracking stylus systems

The usual limitation on speed is the resonant frequency $w_n$ of the pick-up. It is conventional to track at a speed that keeps the frequencies down to about one fifth of the resonant frequency of the pick-up. This is based on sinusoidal signals of value $w$.

However, for the more usual random surfaces and surfaces with sharp edges, the energy transfer characteristics between the surface and the pick-up [7.10]

$$\left[1+\left(\frac{w}{w_n}\right)^2(4\zeta^2-2)+\left(\frac{w}{w_n}\right)^4\right] \tag{7.2}$$

can be optimized by making the damping ratio $\zeta = 0.59$, which has the effect of equalizing the surface reaction up to the resonant frequency. Speeds of traverse, in principle up to five times faster, can be achieved i.e. 5mm/sec. Systems have been tried but repeatability has been a problem [7.11].

The point is that there are still options for designing the mechanics of second order stylus systems. If the damping is neglected, the effective maximum speed of traverse is given by $v$, where $v$ is given for a sinusoidal signal of wavelength $\lambda$, amplitude $A$:

$$w^2 A = R \text{ and } w = \frac{2\pi v}{\lambda} \tag{7.3}$$

where $R$, the reaction on the surface, is the sum of the static and dynamic forces. To increase $w_c$ where lift-off occurs, $R$ needs to be large but as can be seen from the formula there is a square root relationship between $R$ and $w$. So, $R$ has to be greatly increased in order to get an effective increase in $w$. But this increases the chance of damage in the valleys due to the stylus.

If the pick-up has all its weight distributed along the pick-up arm rather than at the end, the value of $w_c$ becomes:

$$w_c^2 A = \frac{3}{m}\left(\frac{R}{l}\right)\cdot\frac{\zeta}{2g} \tag{7.4}$$

and can be doubly increased by increasing $R$ and reducing the value of $m$, the mass of the stylus [7.12].

The idea behind having a fast stylus instrument is not to increase the speed to get a quick profile. It is needed in order to give the possibility of scanning over an area; typical areas would be of a few millimetres squared. Under these circumstances, it is necessary to get a reasonable idea of the x, y positions of the scan so that structure can be measured. This is usually not so important in conventional stylus methods where surface heights are mainly measured.

Other slower ways of getting an areal scan based on stylus methods are available [7.9 (g)]. Sometimes these instruments have dual stylus capability. A 100mm traverse can be offered, which allows straightness as well as roughness to be measured. Stylus methods can be used to scan areas; they may be slower than non-contacting methods but they have better mechanical fidelity.

Also isolating the transducer from the actual pick-up movement by using optical means such as in Figure 7.10 helps to increase the frequency response. Measures such as this are helping achieve higher speeds with nearly constant surface reaction independent of the surface typed. These factors are high-speed fidelity and yet they preserve the debris and film removal attributes of the stylus method.

The next section will consider optical surfaces metrology systems together with a comparison between stylus and optical methods. Some combined schemes will be touched upon.

## References

7.1  Schmalz G. Zeitschrift. V.D.I Vol. 73 p1461 (1929).

7.2  Schmalz G. Technische Oberflaechenkunde Springer, Berlin p66 and 269 (1936).

7.3  Whitehouse D. J. *Stylus Damage Protection Index* Proc. Inst. Mech. Eng. Vol. 214 pt. C p975 (2000).

7.4  Reason R. E. Hopkins M. R. Garrod R. I. *Report on the Measurement of Surface Finish by Stylus Methods*. Rank Organization (1944).

7.5  Whitehouse D. J. *Stylus Contact Method for Surface Metrology in the Ascendancy.* Inst. Meas. Cont. Vol. 131 p48 (1998).

7.6  Whitehouse D. J. *Characterization of Solid Surfaces*. Stylus Methods Ed. Kay G. Larrobee. Plenum.

7.7  Marti O. et al. *Topography and Friction Measurement on Mica Nanotechnology* Vol. 2 p141 (1990).

7.8  Deutshke S. I. Wu S.M. Strolkowsski Int. J. Mach. Tool Des. Res. Vol. 13 p29 (1973).

7.9  Taylor Hobson

    7.9(a) Surtronic Duo 3E 10k SP 0600/Surtronic 3+,8E8KSp07100.

    7.9(b) Form Talysurf intra 2E 20KEPO7/00.

    7.9(c) 120 PGI Form Talysurf Series Z FTSS 6EmCP0800.

    7.9(d) Talycontour TC ZE 3KSP0700.

    7.9(e)Talystep 339-1 in CS 6/95.

    7.9(f) Nanostep 3E 05/98 CP 10K.

    7.9(g) Talyscan 150TS 2E 10K 0400.

7.10 Whitehouse D. J. *A Revised Philosophy of Surface Measuring Systems*. Proc. Inst. Mech. Eng. Vol. 202 No. C3 and 169 (1988).

7.11 Morrison I. *A prototype scanning Stylus Profilometer for Rapid Measurement of Small Surface Area*. Nanotechnology (1994).

7.12 Whitehouse D. J. *Dynamics and Trackability of Stylus Systems*. Proc. Inst. Mech. Eng. Vol. 210 p159 (1996).

# 8

---

# Optical methods

Before discussing specific instruments, it is necessary to consider some of the problems associated with optical methods in the same way that elastic and plastic deformations were examined for the stylus instrument.

Also, it should be realized that the pedigree of optical methods is at least as good as that of the stylus instrument [8.1–8.4]. Just looking at a workpiece from different angles shows how versatile light scatter is.

Figure 8.1 shows a graph on which optical techniques are represented on the abscissa. The ordinate axis has two variables, one being speed and the other fidelity. It is true to say that it is not economically feasible to get both. The closer the method is to the stylus method, the more it is likely to be slower but have more fidelity.

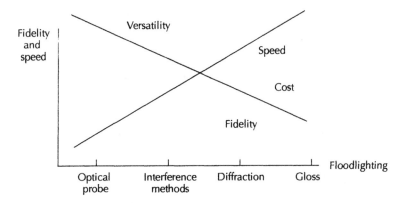

**Figure 8.1** *Optical methods comparison*

Each one of these methods will be discussed briefly. Perhaps the best place to start is with the simple gloss meter. This will be examined after some optical properties have been discussed.

## 8.1   Optical path length

Figure 8.2 illustrates a flat surface upon which two films have been deposited. They are the same thickness but different refractive indices $n_1$ and $n_2$. A stylus instrument measuring this surface would give a straight line because it is responding to the geometry. If an optical probe is used, a clear step occurs where the films change. This is not wrong; it is showing what optics can do under these circumstances.

## 8.2   Optical penetration

Figure 8.3 (a) shows the optical rays penetrating the surface. The amount depends on the conductivity of the surface. It is not negligible on fine nanometric surfaces. It can be of the order of nanometres. Figure 8.3 (b) shows the main problem with optical methods. At a sharp edge or a highly curved peak, secondary scattering results. The surface edge tends to act as a secondary source. This is what makes edges such as grain boundaries so distinctive when viewed under a microscope. In this respect, it is not necessarily a disadvantage because lateral structure is often what is being investigated. However, it is a serious distortion if height information is needed. Figure 8.4 shows what an edge looks like.

## 8.3   Resolution and depth of focus

The resolution of the optical system is dependent on the numerical aperture of the objective lens. In other words, the half angle subtended by the lens as seen from the object. The resolution 'd' is determined by $\dfrac{2\lambda f}{a}$ where $f$ is the focal length of the lens and $a$ is its diameter. This is shown in Figure 8.5. Figure 8.6 shows that the depth of focus $d_1$ is also determined by $\lambda$, the wavelength and $a$ and $f$ as before for resolution. If the object is in air, $\dfrac{a}{2f}$ is the numerical aperture.

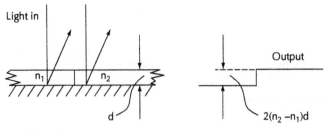

**Figure 8.2**   *Optical path length*

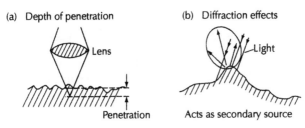

(a)  Depth of penetration

(b)  Diffraction effects

Lens

Light

Penetration

Acts as secondary source

**Figure 8.3**  *Optical problems*

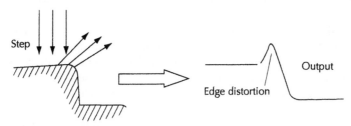

Step

Output

Edge distortion

**Figure 8.4**  *Optical edge enhancement*

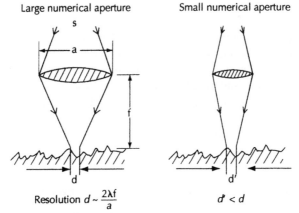

Large numerical aperture

Small numerical aperture

s

a

f

d

d'

Resolution $d \sim \dfrac{2\lambda f}{a}$

$d' < d$

**Figure 8.5**  *Optical probe (resolution in terms of N.A.)*

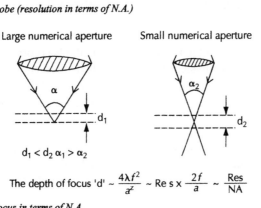

Large numerical aperture

Small numerical aperture

$\alpha$

$\alpha_2$

$d_1$

$d_2$

$d_1 < d_2 \; \alpha_1 > \alpha_2$

The depth of focus 'd' $\sim \dfrac{4\lambda f^2}{a^2} \sim \text{Re s} \times \dfrac{2f}{a} \sim \dfrac{\text{Res}}{\text{NA}}$

**Figure 8.6**  *Depth of focus in terms of N.A.*

Where $\mathrm{Res} \sim \dfrac{2\lambda f}{a}$, in Figure 8.6, $\alpha = f\!\!\Big/_{\!\!a}$. So, if $f\!\!\Big/_{\!\!a}$ has been fixed because of a requirement specification in terms of depth of focus or resolution, it completely fixes $\alpha$. In other words, the ability to measure slope has been fixed by other considerations; $\alpha$ is not a variable.

## 8.4   Comparison between optical and stylus methods [8.5]

The fact that optical methods are rigidly held to optical laws can be an advantage but it can also be a disadvantage. For instance, if the resolution is fixed then the angle subtended by the lens cannot be changed. Hence the angle, the resolution and depth of focus cannot be changed without affecting the others. In stylus methods, it is possible to make the stylus angle and tip dimension independently.

Table 8.1 summarizes the situation regarding the optical and stylus methods. It is clear that, on balance, the stylus methods are best for engineering surfaces, but they do contact the surface.

On some very fine surfaces, there is a considerable difference in value between an optical measurement and a stylus method. Invariably, the optical method gives a larger value than the stylus. This is because the stylus method tends to integrate, whereas the optical method differentiates e.g. it enhances edges. These statements assume that the spot size of both methods is about the same.

If the signal to noise ratio is used as a criterion of merit, they both give about the same result for different reasons. For the stylus, $\dfrac{S}{N}$ is low because $S$ the signal is smaller than it should be. In the optical case, $\dfrac{S}{N}$ is also low but this time the noise $N$ is high. For best results on fine (nanometre) surfaces, the optical and stylus results could well be averaged!

It should be pointed out that neither of the methods is correct or wrong; they are both obeying their own physical laws. Also, when measuring ordinary engineering surfaces, the differences between the methods are small. It is only when very fine surfaces are being measured that the differences are significant. For general references see [8.1–8.4].

These are general comments. Integration by stylus methods is considerably reduced if high resolution styli of 0.1μm are used. Lateral resolution should always be checked before comparisons are made Optical methods tend to have poor lateral resolution.

**Table 8.1**  *Summary of optical and stylus methods [8.2]*

| Stylus | Optical |
|---|---|
| Possible damage | No damage  ✓ |
| Measures geometry  ✓ | Measures optical path |
| Tip dimension and angle independent  ✓ | Tip resolution and angle dependent |
| Stylus can break | Probe cannot be broken  ✓ |
| Insensitive to tilt of workpiece  ✓ | Limited tilt only allowed |
| Relatively slow speed | Can be very fast scan  ✓ |
| Removes unwanted debris and coolant  ✓ | Measures everything good and bad |
| Can be used to measure physical parameters as well as geometry, for example, hardness and friction  ✓ | Only optical path |
| Roughness calibration accepted at all scales  ✓ | Difficult to calibrate by standards |
| Temporal and spatial influence/dynamic effects | Spatial influence/geometric effects |

## 8.5   Gloss meters [8.6]

The **gloss meter** or **scatterometer** is a simple, yet useful, method to assess surface texture. Light is projected onto the surface and the scattered light is picked up by two detectors A and B. A is at the specular angle and B is at some small angle usually at about $10^0$ from it (see Figure 8.7). A quality index $QI$ is formed by taking the ratio $\dfrac{A-B}{A+B} = QI$. If the surface is smooth, all the light enters and $A, B = 0$ and $QI = 1$. A very rough surface scatters all the light evenly so that $A = B$ and $QI = 0$. The index $QI$ is therefore a measure of the smoothness of the surface.

For $R_a \sim \dfrac{\lambda}{8}$ then $QI \sim 1$

For $R_a \sim \lambda$ then $QI \sim \dfrac{1}{2}$

This is a good cheap comparitor. Only surfaces made by a single process can be compared this way. Different processes change the $QI$ in different ways so comparison between processes is meaningless [8.5].

Gloss meters or scatterometers are interesting because they are a practical demonstration of how relevant the metrology unit, in this case the wavelength of light, is to the result. The amount of light scatter hitting detector A depends on the ratio of the roughness (however quantified) and the wavelength of light: the comparison is provided naturally by the laws of physics. It is not often that this comparison is so clear.

A great deal of effort has gone into making this device suitable for in-process gauging. It has not been very successful because of problems with coolant and debris. There are commercial variants using multiple sensors [8.19].

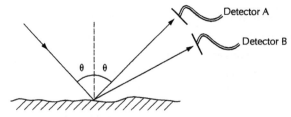

**Figure 8.7**   *The basis of the gloss meter*

## 8.6   Total integrating sphere

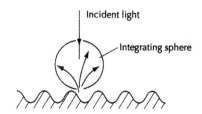

**Figure 8.8**   *The total integrated scattering of an integrating sphere [8.8]*

Figure 8.8 illustrates this technique. All the scattered light is captured in an integrating sphere. The log of the total light is directly proportional to $R_q$ if the surface is Gaussian, which is usually acceptable for fine finishes.

Total integrated scatter $TIS \sim \left( \dfrac{4\pi R_q}{\lambda} \right)^2$ from which $R_q$, the rms roughness, can be estimated [8.7].

## 8.7   Diffractometer

The use of diffraction, in principle, is potentially the best optical method with in-process capability.

This method needs explaining. Usually, a lens is used to magnify the surface. This image is at a specific plane $P_1$. If the diffraction (sometimes called the transform) is required, it is nearer to the lens at $P_2$ [8.9]. This is sensitive to the slopes on the surface rather than the heights (Figure 8.9).

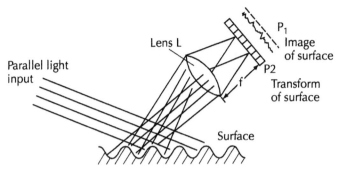

**Figure 8.9**  *Diffractometer*

Spacings of '*d*' on the surface give spectra separated by $D = \dfrac{\lambda f}{d}$ (Figure 8.10) on plane $P_2$. In other words, small detail is magnified more than large detail by the method, which makes it something of a metrological freak. Light scattered at the same angle anywhere on the surface focuses on the detector at one spot. If the surface is relatively smooth i.e. $R_q < \dfrac{\lambda}{6}$ the intensity pattern at the detector is the power spectral density of the surface. This, therefore, could be used to detect tool wear as suggested in Figures 4.5 and 4.6 as well as general machine monitoring [8.1].

This technique has many advantages over conventional optical methods and only one disadvantage.

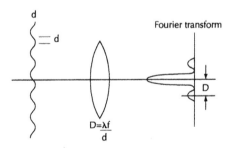

**Figure 8.10**    *The metrological freak diffractometer*

The disadvantage is that the pattern in the detector plane does not look like a surface. It is the Fourier transform of the surface as mentioned earlier but not many users are experts in transforms. This means that the technique tends not be used because the user likes to feel comfortable (by 'comfortable' is meant 'familiar'). If the output pattern on the detector looks odd, the operator wants to know what has gone wrong and obviously how to correct it.

The method is not sensitive to surface speed. The transform is formed at the speed of light. Also, because the ratio of spectral values is used rather than their absolute values, chips, debris or coolant do not affect the results at the detector plane. If the coolant produces a mist then the method fails due to refraction! However, it now looks as if machining of the future will be dry cutting and so no mist is possible. The only rotation in which the method is sensitive is yaw.

From the installation point of view, the technique is good because the position of the light from the surface is insensitive; the light rays are collimated i.e. parallel so that the instrument can be fitted where convenient, providing that the laser is screened from the operator.

## 8.8   Interferometry

Interferometry is usually used for measuring length and is a comparison between two paths of light, as shown very simply in Figure 8.11.

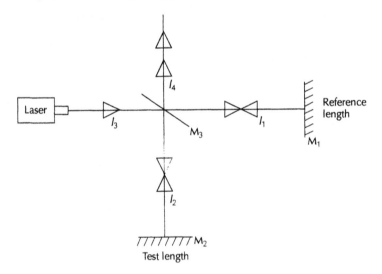

**Figure 8.11**   *Interferometry*

One path, here the reference, is $l_3 + l_1 + l_4$ whereas the test path (that being measured) has a path length $l_3 + l_2 + l_4$ at the detector, therefore the comparison is between $l_1$ and $l_2$. Because the light is coherent, how $l_1$ and $l_2$ combine depends on their phase difference; sometimes adding, sometimes subtracting.

If $M_1$, the mirror, is kept fixed, the length $l_2$ can be determined by counting the number of maxima viewed by the eye. Each maxima corresponds to a path difference of $\frac{\lambda}{2}$ between $l_1$ and $l_2$. Here, the scanning is normal to each of the mirrors $M_1$ and $M_2$. In surface metrology, the situation is different because the comparison is made laterally across the test object. Tolansky [8.10] devised an ingenious method to look at surfaces by interferometry. He placed a flat smooth glass block on the surface with the underside coated. This underside became the reference mirror corresponding to $M_1$ in Figure 8.12 and the surface the test mirror $M_2$. The surface can be considered to be a succession of mirrors across the length of the mirror. The mirror is assumed to have as many reference mirrors making up its length as there are test mirrors. Each pair of mirrors corresponds to the length measuring interferometer shown above. This configuration produces intensity maxima where the optic path difference between the glass block and the surface is in phase, producing a line contour in the xy plane. Also, there can be multiple reflections between each $M_1$ and $M_2$. This enhances the signal and reduces the noise level because the averaging is longer and consequently tends to zero. This is called **multiple reflection**.

Figure 8.12 (c) shows a schematic view and Figure 8.12 (d) shows a way of producing lines of fringes rather than the closed loop contours obtained with the conventional rig. Tolansky simply tilted the reference block. The fringes look like profiles taken across the surfaces. This method only works for relatively rough surfaces and surfaces that are reasonably reflective.

The number of 'mirrors' representing the rough surface in Figure 8.12 (b) is determined by the correlation length of the surface. In general, about six to eight mirrors can be considered for each major waveform. Having more is questionable

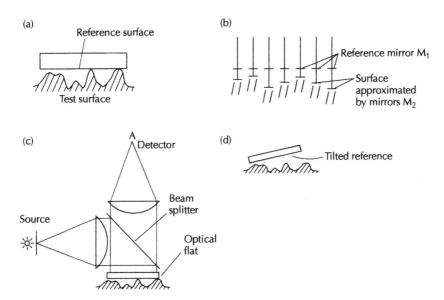

**Figure 8.12**    *The interferometer method for surfaces*

because it is no longer valid to consider the mirrors $M_2$ to be normal to the direction of the light.

The disadvantage of this method is that the test surface has to be the same shape as the reference surface, or at least near to it. Otherwise, too many fringes are formed.

The interferometer has also been used successfully in recent instruments. Basically, the Mireau interferometer is used. The essential optics are shown in Figure 8.13. A spot is focused onto the surface as well as a reference surface near to the objective lens. The system can either be open loop, in which case fringes can be viewed, or closed loop, in which case the instrument becomes an optical follower [8.11, 8.12].

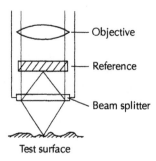

**Figure 8.13**   *The Mireau interferometer*

## 8.9   Optical followers

The original follower was due to Dupuy in 1967 [8.13]. It is a simple method based on the Foucault knife edge method of checking optical wavefronts. In the Dupuy method, a spot illuminates the surface, which is moved laterally relative to the optic axis. There are two detectors positioned behind the back focal plane of the field lens. If the spot hits a peak, it is focused by the optics somewhat nearer to the detectors but past a knife edge positioned at the mean focus. The edge produces an asymmetrical pattern of light on the detectors. This produces two different signals from the two detectors, which are used to drive a servo mechanism. This moves the objective lens upward to re-establish the spot on the surface. Obviously, if a valley rather than a peak comes into the field of the optics, the opposite happens. The movement of the objective lens is taken to be the surface geometry. This method is not too sensitive to the optical properties of the surface but the properties should be consistent. One of the big problems with this original method was the long time taken to complete a scan.

The reason for the slow response is the relatively heavy objective lens in Figure 8.14, which has to follow the surface during the scan. It has been argued that it would be better to move the knife edge to maintain the focus, the knife edge being very small compared with the object lens. Although this is true, there is no benefit because if the objective lens moves $\Delta z$ to keep focus, the knife edge (or equivalent) has to move $(\Delta z)^2$ due to Newton's rule of axial magnification. So, sluggish movement in the case where the lens is moved is replaced by faster but longer movement of the knife edge. The resultant speed is about the same.

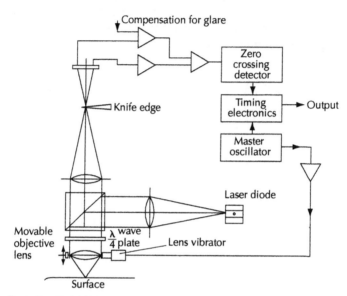

**Figure 8.14** *Dupuy's optical follower*

## 8.10 Heterodyne method

Heterodyne methods were introduced so that a common path could be used for both paths of an interference or focused system. This arrangement considerably reduces the noise in the system. The usual heterodyne arrangement is to have either two light wavelengths or two polarizations [8.14].

A heterodyne follower (Figure 8.15) from NPL has an interesting arrangement, which is in effect similar to the skid system used in stylus instruments. In the case shown, there are two polarizations, one of which is focused and corresponds to the sharp stylus. The out of focus beam produces a much wider spot on the surface so that it cannot reveal detail. This is exactly the skid effect. It has the advantage that the sharp probe can lie in the middle of the skid, something that is not practical in the stylus instrument.

An alternative heterodyne method due to Sommargren [8.15] places the polarized heterodyne paths along side each other rather than on top of each other (Figure 8.16). One of the focused spots is used as a reference and the other as the probe.

The latter describes a circle around the fixed datum spot in order to map the surface. This path is acceptable for flat surfaces but not so good for curved surfaces, although recent developments in software are reducing this problem.

There are many variations on the heterodyne method. Some include Mireau type interferometers as well as using white light [8.16, 8.17]. In commercial instruments there have been many advances in software that enable the areal range to be enlarged using 'stitching' of fields together without losing much fidelity. Also, pattern recognition is routinely used for defect detection.

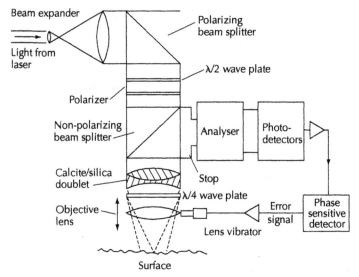

**Figure 8.15**   *The heterodyne follower*

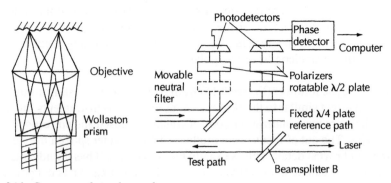

**Figure 8.16**   *Sommargren heterodyne probe*

## 8.11    Other optical methods

One method, which again uses a Mireau type interferometer, does not attempt the optical follower principle. Instead, the whole fringe field is viewed by camera [8.18] and recorded in a computer. The optical system is raised by $\frac{\lambda}{3}$ and the new fringe pattern recorded. This is repeated once more and the contour map of the surface evaluated from the stored fringed pattern. This is a good idea because the areal map is calculated as a whole. The problem with it is that it has to have the optical reference close to the same shape as the object. Curved surfaces can cause problems in concentricity. Also, as will be seen later in cylindricity measurement, care has to be taken when the optical system is stepped axially to get the different fringe maps. Yaw especially is a difficult rotation to avoid. Errors like this or complicated shapes can confuse the computer.

Because of the large number of suppliers of optical scanning and heterodyne interferometry, it is not considered to be constructive to list them here. There are some given in [8.19].

Figure 8.17 shows a confocal system [8.20]. This type of system is used to get very sharp images. It has a strategically placed pin hole near to the detector plane so that only rays entering from the sister optics (relative to the object) are detected. Stray light or widely scattered light is simply ignored. This principle is similar to the early flying spot microscope, in which the detector and the object are scanned synchronously with small spot size. The confocal method is useful for measuring down holes. This situation is prone to reflections and light scatter, which the confocal system reduces considerably.

The confocal method is clever because it absolutely restricts axial light entering the detector except for light in focus. It produces the same signal to noise benefit as did the original flying spot microscope when it was introduced [8.21]. In the flying spot case, the restriction was a light entering the detector from lateral sources. Only the directly illuminated spot contributes to the signal.

There are other well-known optical methods that are not dealt with here. For example, the FECO methods (Fringes of Equal Chromatic Order) and Nomarsky differential methods (Figure 8.18) [8.22, 8.23].

Table 8.2 lists a comparison of some optical and mechanical methods. Optical methods can have greater sensitivities of the limit than conventional tactile methods based on LVDT technology but this extra sensitivity can cause extraneous noise to enter the system. It is here that special optics have to be used, such as in the confocal system.

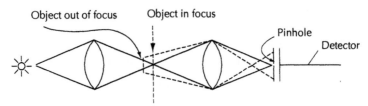

**Figure 8.17**   *A confocal microscope*

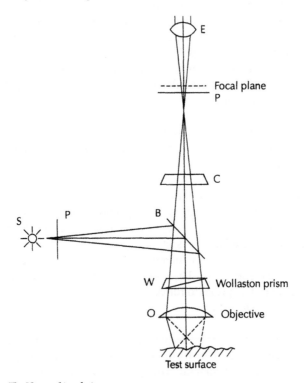

**Figure 8.18**   *The Nomarski technique*

**Table 8.2**   *Properties of metric system for nanometrology*

|  | Resolution (nm) | Accuracy (nm) | Range (nm) | Range/resolution | Max. rate of change nm/s |
|---|---|---|---|---|---|
| Optical heterodyne interferometry | 0.1 | 0.1 | $5 \times 10^7$ | $5 \times 10^8$ | $2.5 \times 10^3$ |
| X-ray interferometry | $5 \times 10^{-3}$ | $10^{-2}$ | $2 \times 10^5$ | $4 \times 10^7$ | $3 \times 10^{-3}$ |
| Optical scales | 1.0 | 5.0 | $5 \times 10^7$ | $5 \times 10^7$ | $10^6$ |

| | Resolution (nm) | Accuracy (nm) | Range (nm) | Range/ resolution | Max. rate of change nm/s |
|---|---|---|---|---|---|
| Inductive Xducers | 0.25 | ... | $10^{4(8)}$ | $2.5 \times 10^5$ | $10^4$ |
| LVDTs | 0.1 | ... | $2.5 \times 10^2$ | $2.5 \times 10^3$ | $\sim 10^4$ |
| Capacitive Xducers | $10^{-3}$ | ... | 25 | $2.5 \times 10^4$ | 10 |
| STM Scales | .05 | .05 | $10^3 - 10^4$ | $2 \times 10^4 - 2 \times 10^5$ | $\sim 10$ |
| Fabry-Perot Etalon (Freq. Tracking) | $10^{-3}$ | $10^{-3}$ | 5 | $5 \times 10^3$ | 5 – 1– |

## 8.12 Conclusions from the comparison of tactile and optical methods

Two typical merit maps are shown in Figures 8.19 and 8.20 and discussed in Section 6 (Fig. 6.8). It is seen that from the engineering point of view, the two methods are complementary. Ideally, the instrument progression should advance along the response axis and the range/resolution axis. There is evidence that in the newer stylus instruments, this trend is happening. It is important to realize that the range to resolution shown here is that along the vertical axis i.e. normal to the surface and not the lateral. Obviously, the lateral resolution depends on the stylus tip size and the range of the traverse of the instrument. These do not pose physical restraints in the same way that optical resolution is limited by the wavelength of light. The specific comparison between stylus and optical followers and similar methods is given in Table 8.1. The figure shows that stylus methods are positioned along the range/resolution axis. This does not take into account in situ measurement with a stylus method. An example is the hand-held stylus instrument, which is positioned near to the machine tool or the object under test if it is very large [8.12 Surtronic].

One point to make is that it is meaningless to speak about correct and incorrect measurements in stylus methods and optical methods. Both techniques have strengths and weaknesses. As a general rule, if the application of the workpiece is non-contacting, such as with an optical mirror, it makes sense to use an optical method to evaluate the surface. If contact is involved, use the tactile stylus method: matching the measuring instrument to the mode of application is a rule of metrology.

Neither method is correct or incorrect. They are different and should be treated as such. It is a serious mistake to try to calibrate optical instruments with tactile standards and vice versa. Both systems should be independently traceable to international standards.

Other methods for measuring surfaces have been omitted here because they lack the versatility and ease of use of the stylus and optical methods. Such methods include capacitance methods, pneumatic methods, ultrasonics and so on. See [5.1] for details.

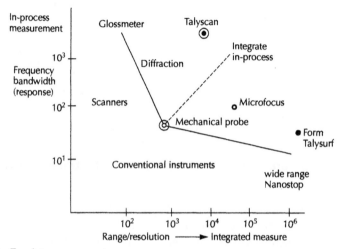

**Figure 8.19**    *Trends in measurement*

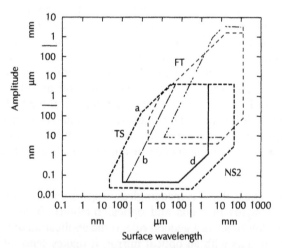

Stylus instruments. TS, Talystep: diamond chisel stylus, solid line; standard stylus, dot-dash line. NS2, Nanosurf 2, long dash line. FT, Form Tallysurf: diamond stylus, short dash line; ball stylus, dash-dot-dot line

**Figure 8.20**    *Surface wavelength vs surface amplitude*

# References

8.1   Whitehouse D. J. *Optical Methods in Surface Metrology*. SPIE Milestone Series Vol. M.S. 129.

8.2   Beckmann P. and Spizzichino A. The *Scattering of Electromagnetic Waves from Rough Surfaces*. Oxford. Pergamen (1963).

8.3   Bass G. G. and Fuchs I. M. Wave *Scattering of Electromagnetic Waves from Rough Surfaces*. New York. Macmillan (1963).

8.4   Ogilvy J. A. *The theory of Wave scattering from Random Rough Surfaces*. Bristol. Hilger (1991).

8.5   Whitehouse D. J. *Stylus contact Method for Surface Metrology in the Ascendancy*. Inst. Meas. And Cont. Vol. 31 p48 (1998).

8.6   Maystre D. *Scattering by Random Rough Surfaces in Electromagnetic Theory*. SPIE Vol. 1029 p123 (1988).

8.7   Bennett J. M. and Mattson L. *Surface roughness and Scattering*. Optical Soc. of America, Washington p24 (1989).

8.8   Whitehouse D. J. *Modern Methods of Assessing the Quality and Function of Surfaces*. Proc. SME paper IQ72 p206 Chicago (1972).

8.9   Rakels J. H. J. Inst. Phys. E Scientific Inst. Vol. 19 p76 (1986).

8.10  Tolansky S. *Application of New Precision Interference Methods to the Study of Topography of Crystal and Metal Surfaces*. Vertrag Tag Dsche. Phys. Ges. 5-7 Gottinghen (1947).

8.11  Delaunay G. Microscope Interferential A. *Mirau pour la Measure du Fini de Surfaces*. Rev. Opt. 32 p610–614 (1953).

8.12  Brodeman R. *Roughness Form and Waviness Measurement by Means of Light Scattering*. Prec. Eng. Vol. 8 p221–226 (1986).

8.13  Dupuy O. *High Precision Optical Profilometer for the Study of Micro geometrical Surface Defects*. Proc. Inst. Mech. Engrs. Vol. 182 pt. 3K p255 (1967).

8.14  Downs M. J. McGivern W. H. and Ferguson H.J. *Optical System for Measuring the Profiles of Super smooth Surfaces*. Pre. Eng. Vol. 7 p211–215 (1985).

8.15  Sommargren G. E. *Optical Heterodyne Profilometry*. Appl. Opt. Vol. 20 p610–618 (1981).

8.16  Zygo Corp. New View 50000.

8.17  UBM Messtecknik. *Microfocus with 200 Series Controllers*.

8.18  Wyko (Veeco) Corporation – Topo range, NT 3300, NT 2000.

8.19  (a) T. M. A. Technologies.
         (b) Zeis.
         (c) Oxford Surface Analytics (Facet).

8.20  Hamilton D. K. and Wilson T. 3D *Surface Measurement using Confocal Scanning Microscope* J. Appl. Phys B27, p211 (1982).

8.21  Young J. Z. and Roberts T. *The Flying Spot Microscope*. Nature Vol. 167 p231 (1951).

8.22  Nomarski G. *Microinterferometre Differentiel a Ondes Polarises*. J. Phys. Rad. Vol. 16 p9–13 (1955).

8.23  Harman J. S. Gordon R. L. and Lessor D. L. *Quantitative Surface Topography Determination by Nomarski Reflection Microscope*. Appl. Opt. Vol. 19 p2998–3009 (1980).

# 9

---

# Scanning microscopes

## 9.1 General

The previous chapter considered optical ways of examining the surface. These are limited by the wavelength of light. As the requirement grows for detail much greater than the wavelength of light can provide e.g. molecularly smooth surfaces, very thin films and their associated step height, it becomes difficult to justify the continued use of optics. One rule of metrology is that the unit of measurement should be close to the size of the feature of interest. In a later section, some methods of using the crystal lattice as a unit will be described but there is another alternative, which is to use electrons instead of light. The equivalent wavelength of a moving electron is $\lambda_e = \sqrt{150V \, / \, eV} \, \text{Å}$ or $\dfrac{h}{p}$, where $h$ is Planck's constant and $p$ *is the electron momentum.* $V$ is the acceleration voltage for the electrons. Putting the relevant values in the formula gives theoretical wavelength for the electron at about 0.1, which is about $10^{-5}$ times the wavelength of light. This practically, is about $10^{-4}$ times smaller.

There are a number of microscopes that make use of electrons rather than light in order to improve the lateral resolution. One uses the electrons in transmission (TEM) and another in **electron scanning mode (SEM)**, which is similar to the flying spot microscope but uses electrons. Because the SEM uses a beam of electrons that has very little spread, the SEM has a very good depth of field because the numerical aperture is very small. These microscopes will not be considered here as their use has been largely superseded by the new generation of microscopes that use a stylus as the probe but on an atomic level. In a sense, the SEM is to optical methods what the new scanning microscopes are to the tactile stylus method.

Another version of the stylus topographic instrument is the **scanning tunnelling microscope (STM)** and the **atomic force microscope (AFM)**. Although new generations of instruments do not appear at first sight to have much in common with the traditional stylus instrument, they do. The only difference basically is that instead of measuring geometry, they measure another feature of the surface such as charge density, force, etc. The basic mechanical system is still second order and more relevantly, the input is still determined by a stylus. Other issues such as the use of skids and references are more relevant to more conventional stylus instruments and basically peripheral to their operation.

Historically, scanning probe microscopy had its origins in a principle first outlined by the British scientist Edward Synge in 1928. He suggested the use of a tiny aperture at the end of a glass tip to raster an illuminated object. This idea was resurrected by J. A. O'Keefe of the US Army Mapping Service in 1956 . To overcome the diffraction limit for microscope studies (i.e. approximately $\lambda/2$), O'Keefe suggested using a tiny aperture in an opaque screen. By placing the aperture close to the surface being studied and scanning the aperture across the surface in a raster pattern, the diffraction limit is effectively overcome. By using this near-field aperture, the resolution could be determined by the size of the hole and not the wavelength of light.

At the time, the required positioning technology to implement Synge/O'Keefe's technology did not exit but now, with the advent of piezoelectric micro-positioning [9.3], this has become possible; the STM uses a mechanical probe rather than a hole and does not use an external source to 'illuminate' the surface. Instead, it uses electrons already present on the surface and the tip; in this case the resolution means that, in theory at least, the STM can have a resolution of an atom. The problem is that of vibration and environmental effects. Also, the tendency of the tunnelling current emanating from one atom to jump to another atom on the same probe makes artefacts very easy to obtain.

The earliest in this family of instruments is the scanning tunnelling microscope. It should be obvious that, because the method relies on an exchange of electrons between the specimen and the tip, the instrument can only be used for conductors or semiconductors. This originally held back the usage of the STM, especially in biology, but recently a range of instruments has emerged, which will be described briefly. In the STM, the probe is actually positioned a distance (albeit small) from the surface and not actually touching (if there is such a thing as touching at this scale). In what follows, some other microscopes will be outlined, after which the STM will be described in more detail. Other details are to be found in [9.3, 9.4].

## 9.2 Scanning microscopes

### The atomic force microscope (AFM) [9.5]

The AFM, which will be described later, can be used to study the topography of non-conductive surfaces. Some versions employ a very sharp tip, usually of fractured diamond mounted on a small cantilevered beam that acts as a spring. Piezoelectric elements move the tip up towards the sample until interatomic forces between the tip and the sample deflect the cantilever. The AFM monitors the amount of deflection (using optical interferometry, reflected beam or tunnelling methods) to sense the amount of force acting on the tip, with the STM used to monitor the cantilever movement. The AFM enables most specimens to be examined.

### Laser force microscope (LFM)

The LFM uses a tungsten probe having a very fine tip. Piezoelectric controls at the base of the probe vibrate the tip at just above its mechanical resonance. The tip is moved to within a few nanometres of the surface. Weak attractive forces from the sample reduce the resonance frequency of the wire, reducing its vibration amplitude by effectively changing the spring rate. The amplitude of the vibration is measured. There should be no contact at all with the surface. This instrument is primarily used for imaging microcircuits.

### Magnetic force microscope (MFM)

Much like the LFM, the MFM uses the probe vibrating at its resonance frequency. The difference between the MFM and LFM is that the probe is magnetized. As a result, this type of microscope senses the magnetic characteristics of the sample. The MFM modulates the resonance frequency as in the case of the LFM. Its main use has been for magnetic media like recording heads etc.

### Electrostatic force microscope (EFM)

Similar to the LFM and MFM, except that the probe has an electric charge, the EFM has been used most to detect the surfaces of electrically doped silicon for use in microcircuits.

### Scanning thermal microscope (SThM)

The probe here is designed as a thermocouple. This is a tungsten wire with a tungsten-nickel junction. The tip voltage is proportional to the temperature. The tip is heated by an applied current and then positioned near to the sample surface. Heat lost to the sample, which varies according to the space between the probe and the surface, is a measure of the gap and hence the topography. This is used to map temperature variations.

### Scanning ion conductance microscope (SICM)

The SICM uses a micro-pipette probe containing an electrode. It is intended for use in measuring the activity in living cells.

## Near field scanning optical microscope (NSOM)

This is the nearest device to that suggested by Synge. This instrument scans a sub-micrometre aperture across and very close to the sample. The NSOM emits light to the sample and measures the modulation produced by the sample as the aperture is scanned across the surface.

All the microscopes in the STM family have two things in common: they have a probe except for the NSOM, which has a hole, and they have a requirement for very accurate micro-positioning. This latter requirement is relatively new in surface metrology. Conventionally, the resolution in the x direction has been much coarser than that for the z direction, reflecting the emphasis on height information. It is only when atomic- or molecular-sized objects are being examined that lateral magnifications for the probe type instruments need to be of the same order.

## 9.3    Operation of the STM

The operation and theory of the STM will be outlined briefly below.

Information from the STM is obtained in two ways. One is a null method; the other is open loop. These are described in 9.3.1 and 9.3.2.

### 9.3.1    Constant current mode

As Figure 9.1 shows, the tip is moved across the surface. The distance between the tip and sample is maintained by keeping the tunnelling current constant. For any change in current (say, an increase) the STM compensates z, withdrawing the tip slightly and vice versa. The output from the STM is the voltage applied to the tip piezoelectric control mechanism to maintain a constant current.

### 9.3.2    Constant z (Figure 9.1)

In this, the STM keeps the tip at a constant average $z$. The STM then measures the change in the current. This is the output.

In Figure 9.2, the output is a smoother curve but because the servo system is constantly making height corrections throughout the measurement, it is quite slow. However, because it is acting as a 'follower', this mode tends to be good for an irregular surface where the constant-height method would be too risky or jumping occurs (Figure 9.1).The latter is preferred for smooth surfaces, which can be measured more quickly.

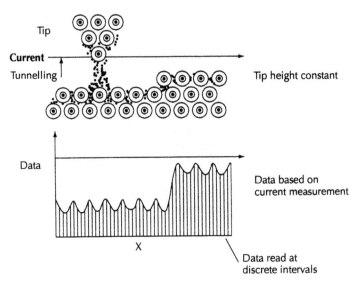

**Figure 9.1**    *Constant height mode scanning*

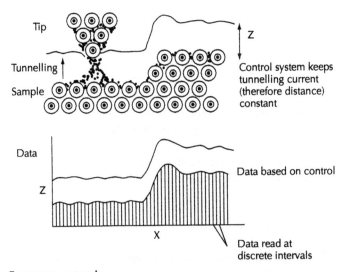

**Figure 9.2**    *Constant current mode*

What does the STM measure? According to quantum mechanics, electrons on a surface behave both as particles and waves (probability density). One result of this is that electrons can be thought of as a cloud above the surface, as in Figure 9.3. When two surfaces or, strictly, electrodes are brought together there is a probability that electrons will pass through the potential barrier, which is trying to inhibit such movement i.e. the work function. As the surfaces come closer together, this probability gets higher until, when the gap is sub-nanometric, a measurable current will exist. This tunnelling $J_T$ is given by $J_T(z)$.

$$J_T(z) = J_T \alpha \exp\left(-A\psi^{1/2} z\right) \tag{9.1}$$

where $m$ is the free-electron mass, $\psi$ is the barrier height and $z$ is the spacing. With barrier height $\psi$ (the order of a few electron volts) a change of $z$, the separation by a single atomic step of two to three changes the current by up to three orders of magnitude. This phenomenal increase in the sensitivity of the current-to-gap makes the family of scanning microscopes possible: there has to be significant change in current for a gap change of Ångstroms, otherwise the atomic distances would not be measurable.

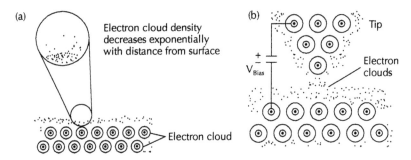

**Figure 9.3**   *Electron clouds between surface and tip*

Using only the spacing dependence given in equation (9.1) and a spherical tip of radius $R$, the lateral spread of a surface step is about $3(2R/A\psi^{1/2})^{1/2}$, which implies that to resolve 10nm requires a stylus with a radius of the same dimension (using a basic rule of metrology).

In the constant-current mode of $J_T$ this implies that $\psi^{1/2} z$ = constant. Thus the $z$ displacement as shown in Figure 9.2 gives the surface topography only for constant $\psi$ – the work function. So, unfortunately, if there is a change in work function on a geometrically featureless surface, it will register as a topographic feature. One possible way to differentiate between changes in topography and work function is to modulate the gap $z$ while scanning at a frequency higher than the cut-off of the control system, but even then there are complexities.

The tunnelling current of electrons, to a first approximation, can therefore describe the topography of the surface – showing the actual position of the atoms. In passing, it should also be pointed out that there is a finite probability that atoms tunnel as well as electrons. This is important when considering the use of the STM as a low-energy means of moving atomic material. The relationship to topography should, however, always be remembered for what it is, a measure of electron state densities. If the bias voltage is changed, then the so-called topography will also change. Measurement of state density indicates the number of electrons either present or allowed at the particular energy determined by the bias voltage. It is obvious that, rather than being a disadvantage, this can be turned to an advantage. The current can provide more than topography. By exploring the effect of bias details, which might be important in chemical bonding, chemical composition and even crystalline structure can be obtained.

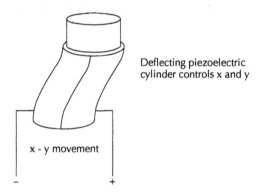

Deflecting piezoelectric cylinder controls x and y

x - y movement

−          +

**Figure 9.4**   *Movement of probe, typical arrangement*

### 9.3.3   *Micro-positioning*

To achieve nanometric positioning and control, the whole system has to be stable. There are a number of ways of achieving this. The simplest of all is to make the system small. This has a number of beneficial effects. One is that the mechanical loop between the stylus and the specimen is small, which makes it more difficult for vibrations to influence any one arm of the loop. It does not matter if the whole loop is affected. Also, the smaller the units making up the mechanical loop, such as screws, levers, springs, etc, the higher the resonant frequency and the less chance there will be of it being excited by motors, gearboxes, drives, etc. They are also less likely to interact with each other. One reason why piezoelectric elements are used is because of their very high stiffness.

Piezoelectric devices can move x y stages to the precision required in STM and AFM instruments but, unfortunately, they suffer from non-linearity and hysterisis. They may be stable but an external sensor is needed to find out where the stage is and to control the position by closing the metrology loop.

The sensor used by Queensgate is the capacitance micrometer. These are small and simple devices with a very high resolution [9.6].

Piezoelectric devices are usually large with respect to the range of movement possible. For example, in Figure 9.4, the actuator tube has to be 100mm long to achieve 100μm range: a factor of $1:10^3$. It is suggested [9.3] that because the stiffness of the piezoelectric actuator is very high, some of this can be sacrificed to reduce the size/range ratio. A factor of 10 is possible but it makes the effective stiffness that much lower, which can cause resonance problems.

At the same time, the mechanical loop should have a high resonant frequency by being small and stiff. So any suspension system for the instrument should have a low one so that it can best 'short-circuit' environmental and instrumental low-vibration frequencies.

It is also advantageous to keep the whole system simple. The simpler it is, the less likely it is that interaction can take place and the more chance there is of maintaining integrity of shape.

Also, if the mechanical loop is small, the chance of getting temperature differentials across it is small; the whole system moves together in temperature. This leads on to the important point about thermal coefficients. The important point is that it does not matter if the elements of the mechanical loop have high or low conductivity; providing that it is the same across the loop. Then temperature does not cause the loop to change shape. It is the shape change that causes the signal to drift, that is if one arm becomes longer than the other in the effective calliper. For many reasons mentioned earlier, the loop should be as symmetrical as possible. Skid considerations are not applied to STMs, or AFMs; the property being monitored is different. Also, skids are used for vertical integration and should not be concerned with lateral positional accuracy.

## 9.4   The atomic force microscope

Whereas the STM actually traces profiles of constant charge density at a particular value of energy determined by the bias voltage, the atomic force microscope measures atomic force. In essence, it comprises a stylus, a cantilever and a means of measuring the deflection of the cantilever tip. The stylus is at the end of the cantilever. Atomic forces at the surface attract the stylus when it is very close. These forces are balanced by the elastic force generated by bending the cantilever. The bend of the cantilever when stationary is therefore a direct measure of the atomic force. There is another mode in which the beam is resonated; the atomic forces then effectively modify the spring rate of the cantilever. The problem arises of how to measure the deflection. One method is to use an STM system riding on the cantilever as shown in Figure 9.5.

Alternatively, an optical readout can be obtained by illuminating the top of the probe and monitoring the deflection of the beam of light (Figure 9.6).

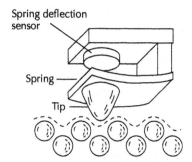

**Figure 9.5** *Atomic force probe using STM measurement system*

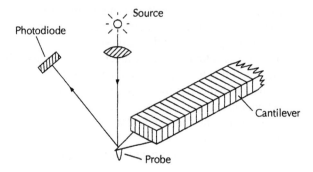

**Figure 9.6** *Optical readout of cantilever angle*

## 9.5 Scanning microscopes: conclusions

The new scanning microscopes outlined above are very interesting developments. There is no reason to doubt that they will continue to develop at the same rate. The difficulty is calibration and deciding exactly what is being measured. They continue to produce astonishing pictures but, apart from structure, the measurements are somewhat less than objective. For example, is there such a thing as 'touching' at these atomic levels? A balance of forces is a better description. Should conventional stylus methods be addressed similarly?

The advent of scanning microscopes based on quantum effects has added a new dimension to surface metrology. Whereas in the past dimension and position have been isolated from surface effects, it is now clear that dimension, position, tolerance and surface metrology overlap in range, indicating that new surface metrology instruments will have to have the capability to measure all the above features. Furthermore, at present, it seems that this integrated instrument or at least its measuring loop will be small, being of the order of centimetres in size rather than metres. Also, there has to be a rethink of the manufacturing process and the machine tool. Their applicability has to be able to encompass the sub-micro range in all of the features mentioned above. One of the key pointers of instrument design (and incidentally of machine tools) is their ability to help develop new generations of the same instrument.

Scanning probe instruments can help to miniaturize the new generation of probes, which are likely to turn into new machines of assembly. One rule of metrology is that the unit of measurement should be near to the size of the feature being assessed. The same logic can be carried to the instrument size.

Perhaps the most distinctive difference between conventional surface measurement and scanning microscope usage is the importance of shape, form and waviness. These represent a large problem in engineering surface metrology as can be seen from the earlier part of this book. They are rarely, if ever, mentioned with respect to scanning microscopes. The reason for this is that in the area involved i.e. square mm (or smaller), curvatures due to shape and form are very small and are ignored. Scanning microscopes are largely areal in application and not three-dimensional. Engineering surface metrology deals with the totality of the piece.

As the size of motors and actuators shrink because of miniaturization, the presence of the longer wavelengths will start to impinge on the scale of the scanning microscopes. It will be interesting to see how the metrology develops. There is no capability at present.

## 9.6  Instruments 'horns of metrology': conclusions [9.7]

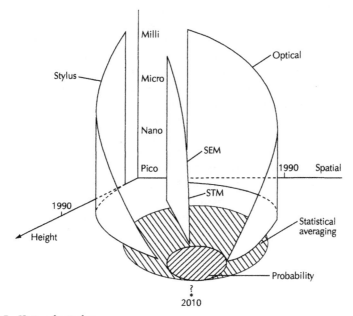

**Figure 9.7** *Horns of metrology*

Figure 9.7 represents the trend in the development of surface metrology instruments. Stylus instruments originally were only concerned with height (z), reflecting the tribological need to measure surface height, whilst optical and other microscopes tended to measure spatial detail (x and y). Over time, both types have recognized the need for measuring the complementary axes as well as the preferred axis. So, stylus methods have been developing areal scan techniques to augment the height and vice versa. The result is a convergence of both types to each other at the nanometric scale of size.

It seems likely that there is scope for a new type of instrument that gives equal weighting to all axes.

There are other surface measuring instruments, which are well known in industry. Pneumatic methods and capacitative techniques have been used since the 1930s. They have not, however, been as useful as the stylus, optical and now the scanning microscopes. They have therefore been left out of this discussion. This is not to say that they do not have a place in metrology. Capacitative probes in particular have been used extensively for position sensing. Some information on these is given in [9.7].

In summary, there is another more profound change that is occurring as the scale of size decreases. This is concerned with the way in which the signal is presented. In normal engineering, scales corresponding to 'milli' and 'micro' (in Figure 9.7) positions of stylus or probe are deterministic i.e. the scale of size of the movement or object position is the same or bigger than the probe size itself. As the scale reduces to 'nano', the probe position is achieved in any axis by having the probe size larger than the movement scale of size or position so that statistical spatial average can determine the signal. This is why sub-nanometre positional accuracy can be achieved on any *single* axis despite using relatively large probes. This spatial integration is not possible on the atomic scale if nanometre accuracy is to be achieved on all axes, which is the aim of the $M^3$ instrument. Here, spatial averaging cannot be achieved on a $nm^3$ scale. The result is that the signal is determined purely by quantum effects, which can only be made workable by integrating in time i.e. allowing the signal (number of electrons or photons) to build up to a measurable level. This time delay requires that the instrument has to have stringent environmental insulation, which may not be achievable in practice. So, what was an engineering or metallurgist problem a few years ago is now turning into a physicist's problem of measurement. The problem is that the engineer is still expected to be able to deal with it!

# References

9.1   Quate C. F. *Scanned Probe Microscopes*. AIP Conf. Scanned Probe Microscopy Ed. H. K. Wickramasinghe. Santa Barbara (1991).

9.2   Abbe E Archiv Microskopische Anat, 9413 (1873).

9.3   Binnig G. and Rohrer H, Gerber, C. and Weibel E. Phys. Rev Lett, 49 p57 (1982).

9.4   Binnig H. and Rohrer H. *The Scanning Tunnelling Microscope*. Surf. Sci. Vol. 152/153 p17 (1985).

9.5   Martin Y. Williams C. G. and Wickramasinghe H. K. *Atomic Force Microscope Force Mapping and Profiling on a Sub 100 Scale*. J. Appl. Phys. Vol. 61 p4723–4729 (1987).

9.6   Hicks T. R. and Atherton P. B. *Nanopositioning Book*. Queensgate Instruments Ltd. (1997).

9.7   Whitehouse D. J. *Handbook of Surface Metrology*. IOP Publishing. Bristol and Philadelphia (1994).

# 10

---

# Errors of form (excluding axes of rotation)

## 10.1 General statement

One of the biggest problems of form is concerned with one of its boundaries i.e. waviness. Waviness is difficult to deal with from a purely geometric standpoint. It is not a measure of a metal-air boundary as is roughness. Neither is it a deviation from a perfect Euclidean shape as are the measurements of the deviations from straightness, flatness and roundness (Figures 10.1 and 10.2). At least, in these latter cases, there is some defined perfection. (The definition can be written down formally.) This makes the measurement problem easier. On the other hand, the wide range of different engineering shapes that have to be contended with is considerable. Some general classification will be discussed in what follows.

To be consistent with what has already been said, deviations from, or concerned with, a linear variable will be considered in this chapter. Form can be considered to extend into straightness and then flatness. Geometric shapes concerned with curves such as circles, epi-trochoids etc. are examined later in Section 11.

Errors of form suffer from the opposite of roughness. In roughness, the short wavelengths are known to be subject to instrumental constraints such as stylus dimension. It is the long-wavelength boundary of waviness that is difficult to define. This is the short wavelength boundary of form. In form errors, the long wavelengths are determined by the ideal shape specified e.g. a circle and it is the short-wavelength boundary with waviness that has to be specified. Because of the difficulty of defining anything in absolute terms, it is usual to define the short wavelength of form in terms of the size of the piece. The short limiting wavelength of form is usually defined as a fraction of the relevant dimension of the workpiece. A factor of one-third or a quarter has been used. Wavelengths less than this are difficult to explain in terms of errors of form.

Errors of form are relatively easy to characterize: they are broken down into Euclidean shapes such as circles, planes etc. This is easy when compared with the problem of characterizing roughness. However, complex problems arise in the methods of qualifying form error. In what follows, emphasis will therefore be placed on the assessment problem.

The ideal form itself can be regarded as a skin in space. It needs a certain minimum number of points to describe its shape and position.

The ideal skin so defined is infinitesimally thin but practical surfaces are not. If zonal methods are being used in assessment, such as a minimum zone in which the thickness is a measure of the minimum peak-to-valley distance from the skin, one more data point is needed to fix the thickness of the skin. That is, the minimum zone sphere needs five points of constraints corresponding to three points for origin, one for size (a radius) and one for thickness of zone (see Table 10.1).

Errors of straightness are often due to errors in machining and slideway error, but they can also be the result of sagging of the workpiece under its own weight, thermal effects produced during machining, stress relief after machining and many other reasons. The types of component usually involved in this sort of assessment are shafts, slideways etc.

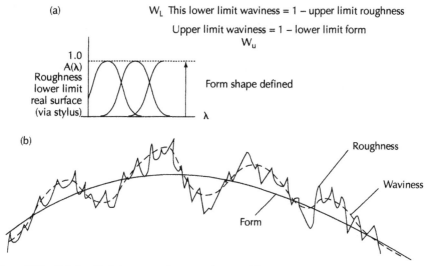

**Figure 10.1**   *Relationship between roughness, waviness, and form*

The boundaries of waviness $W_L$ and $W_u$ are both defined relative to arbitrary upper limit roughness and lower limit form. This makes characterization very difficult for waviness and, to a lesser extent, form.

The only practical way is to relate the geometry to cause and use as in Figure 10.2.

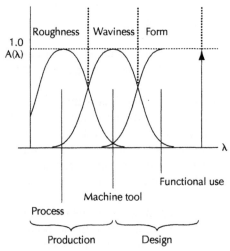

**Figure 10.2**  *Quality regimes for surface geometry*

There are, as in most metrological problems, three separate difficulties that need to be considered. The first is the nature of the problem, the second is the method of assessment and the third is the method of display.

Table 10.1 gives a recent version of the symbols for the various components of form error as shown.

**Table 10.1**  *Error of form symbols*

| | | | |
|---|---|---|---|
| ◯ | Roundness | — | Straightness |
| ◉ | Concentricity | \ / | Co-axiality |
| ‖ | Parallelism | ⟕ | Cylindricity |
| ▱ | Flatness | ✐ | Runout |
| ⟂ | Squareness | | |

Table 10.2 shows the number of points to define the shape of the geometry.

**Table 10.2**   *Constraints on form geometry*

| Feature | No. of Points |
|---------|---------------|
| Line | 2 |
| Plane | 3 |
| Circle | 3 |
| Sphere | 4 |
| Cylinder | 5 |
| Cone | 6 |

The points in the column do not refer to degrees of freedom. For example, one degree of freedom i.e. a translation in the x direction is a movement along a line, which needs two points to establish it. Similarly, a rotation about one axis is one degree of freedom but a circle requires three points to establish it. The points are geometrical constraints imposed by the shape. It is useful to know these points because it enables unambiguous paradigms to be developed, as will be seen.

Table 10.3 (overleaf) shows how Table 10.2 has to be modified to take into account the assessment of the deviation from the shape [10.2].

## 10.2   Straightness and related topics

The terms roundness and straightness are somewhat misleading. They should really be 'the departures from true roundness' and the same for straightness. Common usage has reduced them to the single word.

Take straightness, for example. The test surface has to be compared with a true line somewhere in space. The first problem is generating the line. The same is true for roundness. There is, however, a significant difference between straightness and roundness. It is much easier to generate a circle in space than it is a line; the perfect circle is continuous but the line is not. There is a certain amount of support from all around the bearing when the perfect circle is being generated. This does not occur in straightness.

## 10.3   Measurement

In measuring straightness, it is common practice to measure the error by methods that take account of the length of the surface to be measured. For example, if a

**Table 10.3**   *Zonal points*

| Function 2D Line | Figure | Points for definition | |
|---|---|---|---|
| | | 2 | |
| Minimum deviation from line | | 3 | |
| Plane | | 3 | |
| Minimum deviation from plane | | 4 | |
| Circle | | 3 | |
| Maximum deviation from minimum circumscribed circle (ring gauge) | | 4 | |
| Maximum deviation from maximum inscribed circle (plug gauge) | | 4 | |
| Minimum zone | | 4 | |
| Function | Figure | Points | Deviations from |
| Sphere | | 4 | 5 |
| Cylinder | | 4 | 5 |
| Cone | | 5 | 6 |

surface plate is to be measured for flatness, the instrument chosen would be such that it would ignore scratches and scraping marks and detect only the longer wavelengths on the surface. However, if the workpiece is the spindle of a watch component only a few millimetres long, then a different approach is required since the spacings of the undulations that could be classed as errors of form are now very small and it becomes much more difficult to separate form error from the surface texture. This problem is typical of nano metrology.

Because the measurement of the straightness of small components to a high order of accuracy is fairly easily achieved by the use of instruments having accurate slideways, the graphical representation is usually satisfactorily displayed on a rectilinear chart.

In using such instruments, the parallelism or taper of opposite sides of a bore or shaft can be made simply by taking multiple traces on the recorder. Note that this is correct, providing that there is no relative movement between the workpiece and the datum when taking the traces.

As has been stated, it is difficult to isolate the method of assessment of a parameter from the method of measurement. Straightness measurement can be achieved in a number of ways, either directly or indirectly. The direct method involves comparing the surface with a straight line datum, which can be either a mechanical reference or the line of sight of a telescope or a laser beam. The indirect method is one in which either the local slopes or the local curvature is measured across the surface. The slope or the curvature can be measured and the straightness obtained by integration. As a first step in characterization, the data has to be provided. Then characterization proper can take place usually by fitting a suitable function to the graph.

All of these methods ultimately must end up with a graph that shows the deviations of the surface from a straight line reference somewhere in space as in Figure 10.1. In general, the reference is not in the same direction as the surface; neither is it within the measured profile [10.1].

The obvious way to measure straightness involves measuring the test piece relative to a reference axis. The problem is that of generating a straight line to use as the reference axis.

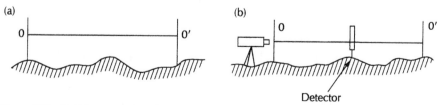

**Figure 10.3**   *Straightness measurement: the direct approach*

Figure 10.3 shows the direct method. Today it is reasonable to use this method because of the availability of lasers. The deviations are measured relative to a light sensitive quadrant detector. The reference $0,0^1$ is set up at the start by adjusting height at 0 and $0^1$.

Even a laser has its drawbacks. One is that it has to be accompanied by health warnings. Another is that it has quite a spread on the detector, as seen in Figure 10.4 (b).

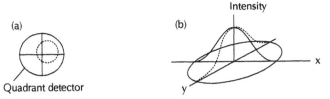

**Figure 10.4**   *Centre of gravity of laser beam*

If the laser spot is central, the output from all quadrants is the same but if it is off-centre (Figure 10.4 (a)) an output error signal is generated by the detector: each quarter develops a signal different from its opposite. This is interpreted as a mis-alignment or deviation from the axis. A laser need not be used – an alignment tele-scope has been used for years – in fact since the 1930s. The misalignment at each position is the straightness or, more precisely, error from the line of sight (Figure 10.4 (a)). A typical alignment telescope is shown in Plate VIII.

One problem with lasers, which is usually easy to compensate for, is the fact that the intensity pattern across the beam is Gaussian i.e. bell-like and not uniform, so that finding the centre of the beam on the detector requires estimating the centre of gravity of the intensity position curve in two dimensions (Figure 10.4 (b)).

It should have a typical specification as follows.

(a)  Micrometer range ± 1.2mm.

(b)  Optical axis parallel to mechanical to 3 arc sec.

(c)  Concentric to 6μm.

(d)  Field of view from 50mm (2 inches) at 2m (6.5 ft.) to 600mm (24 inches) at 30m (100ft.).

(e)  Focusing range 25mm (1 inch) to infinity.

(f)  Accuracy 0.05mm (0.002 inches) at 30m (100ft) and proportionately for longer or shorter distances.

Figures 10.7 (b) and 10.7 (c) show the use of the alignment telescope in squareness and flatness, to be discussed later (Figure 10.10 (a) is for straightness).

If an engineer's flat is available, then bars or straight edges can be checked using a very elegant method involving slip gauges and what is called the **wedge technique**. Plate VIII shows the system as supported (see also Figures 10.5 and 10.6).

The test bar is supported at the Airy points i.e. 0.554$L$ apart when the bar is of length $L$. The clever bit is to put slip gauges of unequal height at the two support points. Subdivide the space between the supports into equal distances – corresponding to the width of the gauge. Then literally build up the wedge across the supported length with gauges. The actual deviation is taken at the corner.

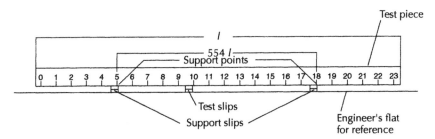

**Figure 10.5**  *Use of engineer's flat*

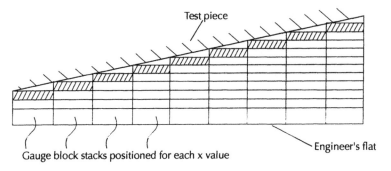

**Figure 10.6**  *Wedge method for straightness [10.1]*

In order to avoid the problem of generating a true line, the indirect method of measuring angles is sometimes used. Here, the length is split into equal intervals and the angle measured by a clinometer or autocollimator. The principle of the autocollimator is shown in Figures 10.8 and 10.9. Figure 10.8 (a) shows a focus by the objective lens of the autocollimator at point 0. If the mirror (or test piece) is tilted, then the focus point moves accordingly on the detector to the image I by an amount 'a'. Knowing the focal length f and the measured 'a', the tilt of the object (in this case the mirror) can be found. A typical autocollimator is shown in Plate IX (a). Table 10.4 shows the typical range of autocollimators.

Figure 10.7 (a) shows how the angle is measured incrementally along what could be a lathe bed, for example.

Figure 10.7 (c) is measuring flatness as seen later. Alignment telescopes have been used extensively in industry.

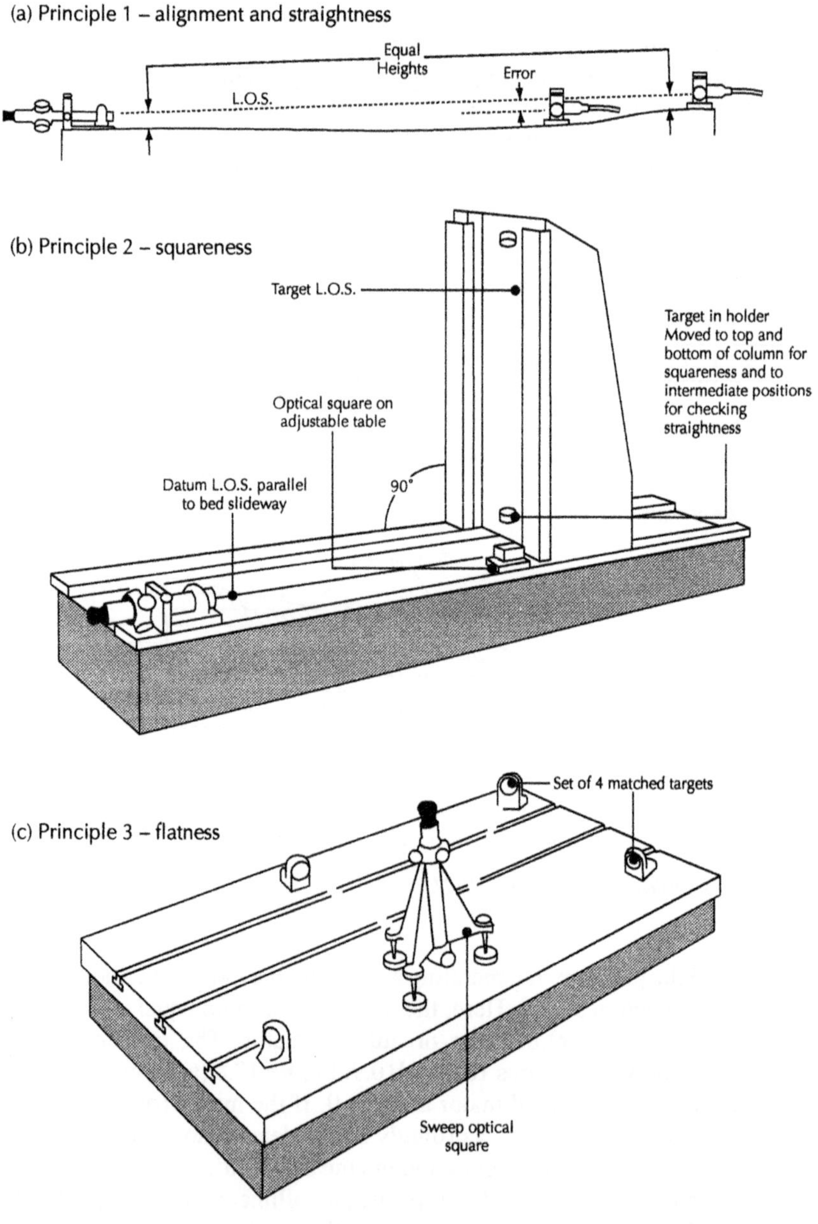

(a) Principle 1 – alignment and straightness

(b) Principle 2 – squareness

(c) Principle 3 – flatness

**Figure 10.7**  *Uses of alignment telescope*

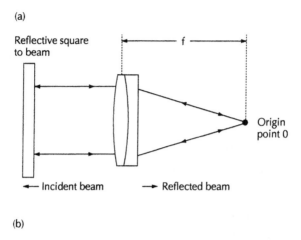

(a)

Reflective square
to beam

f

Origin
point 0

← Incident beam     → Reflected beam

(b)

Reflector
tilted

→8←

28

f

Image 1

a

Origin
point C

**Figure 10.8** *Autocollimator principle*

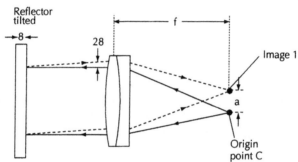

(a)

Autocollimator

Mounted reflector on

Surface being checked

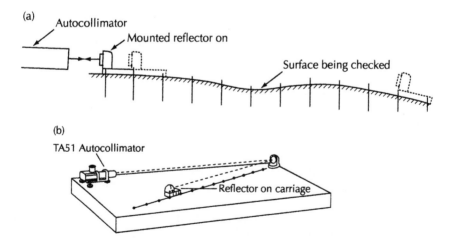

(b)

TA51 Autocollimator

Reflector on carriage

**Figure 10.9** *Measuring straightness using (a) angular method and (b) flatness*

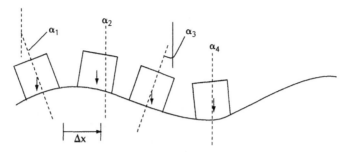

**Figure 10.10**  *Use of clinometer for straightness*

In Figure 10.10 (a) tilt device is used to measure angle $\alpha_1$, $\alpha_2$, etc. The angle is usually taken at equal intervals $\Delta x$, which is the width of the angle device. The height at any $x_i$ is $a_i$,

$$\text{where } a_i = \Delta x \sum_{j=1}^{N} \alpha_j$$

Angle measuring instruments for measuring straightness and flatness are not confined to autocollimators. With electronic clinometers such as the Talyvel (Figure 10.11), the availability of an electronic output to what is essentially a plumb line facilitates easy angle measurement.

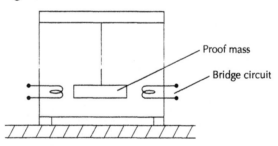

**Figure 10.11**  *Schematic of electronic spirit level*

If the proof mass in Figure 10.11 is not symmetrical within the case due to a tilt at the feet, the electronic bridge gives an output proportional to the tilt.

Notice that although angle measurement is not regarded as surface metrology, it can be used as a tool of surface metrology. The same is true for distance and position. It is a statement of the obvious that all engineering metrology is entwined.

Plate IX (b) shows an autocollimator in use measuring the angles on a reference polygon.

**Table 10.4**    *Typical range of autocollimators*

| Accuracy over 1 min of arc ++ | sec | 6 | 1 | 0.5 | 0.++ | 0.2++ |
|---|---|---|---|---|---|---|
| Accuracy over total range | sec | 30 | 1 | 2 | 0.2 | 4 |
| Range of measurement | min | 60x60 | 900 | 10 | ±20 | ±400 |
| | sec | – | (±1500 ext) | – | | |
| Direct reading to | sec | 60 | 0.5 | 0.2 | 0.01 | 0.1 |
| Working distance for full measuring range | m | 0.5 | 1 | 9 | 5 | 5 |
| | ft | 1.5 | 3 | 30 | 15 | 15 |
| Visual setting range (field of view) | min | 180 | 38 | 19 | 30 | 210 |
| Readout means | | Graticule | Micrometer & graticule | Micrometer | Digital display | Digital display |
| Measurement axes | | 2 | 2 | 2 | 2 | 2 |
| Light source for measurement | | 6V 2 Watts Lamp | 6V 2 Watts Lamp | 6V 2 Watts Lamp | Infra-red LED | Infra-red LED |
| Light source for viewing | | – | – | – | Yellow LED | Yellow LED |

It is possible to get rid of the arbitrary tilt by measuring curvature incrementally and integrating twice. An apparatus for doing this is shown in Figure 10.12. This uses four probes [10.11].

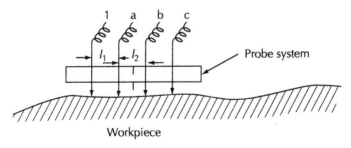

**Figure 10.12**    *Measuring curvature with multiprobes*

## 10.4    Assessment and classification of straightness

There is one serious problem arising from the use of indirect methods. This is the enhancement of noise. Methods such as the autocollimator or spirit level are basically differentiation methods that reduce the signal to noise ratio and whilst this is not too serious a problem in the measurement of straightness, it can be very serious in measuring flatness. Methods have to be adapted to reduce noise. This is a good example of the balance that has to be achieved between ease of measurement and complexity in processing. Luckily, the latter is becoming less expensive and faster so that clever spatial configurations for probes, for example, can be utilized more often.

Assuming that such a data set as seen in Figure 10.13 has been obtained by the appropriate means, the problem of characterizing it remains.

A proposed way has been to use a best-fit least-squares line drawn through the profile to establish the reference. Then, from this line, the sum of the maximum positive and negative errors is taken as a measure of the departure from true straightness.

Deriving such a line is equivalent to fitting a first-order regression line through the data set representing the profile of the workpiece, as in fitting reference lines to separate waviness from roughness, although here the subtle difference is that the drawn line is an end in itself. Deviations from it will be the deviations from the intended shape. Let the vertical deviation from an arbitrary level at distance $x$ be $z$ and for all $x$ let $Z$ be the value of the best fit height at $x$.

$$S = \int (z - Z)^2 (\cos^2 \alpha) \tag{10.1}$$

The problem is that of minimizing $S$ over all $x$, thereby giving equations for the slope and the intercept of the best fit line.

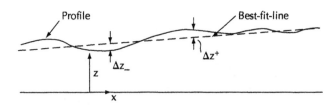

**Figure 10.13**   *Straightness – best-fit line*

$$2\alpha = \tan^{-1} .2 \left| \frac{\sum \sum x_i z_i - (1/N)\sum_{i=1}^{N} x_i \sum_{i=1}^{Nz_i} z_i}{\sum \left(x_i^2 - (1/N)\left(\sum x_i\right)^2\right) - \sum_{i=1}^{N} \left(z_i^2 - (1/NM)\left(\sum z_i\right)^2\right)} \right| \tag{10.2}$$

because the angles are small, this reduces to:

$$m = \frac{\sum x \sum z - N \sum xz}{\left[\sum x\right]^2 - N \sum x^2} \qquad (10.3)$$

However, these small angles will not be valid for microminiature parts, as they are evaluated with small-range coordinate-measuring machines, and so:

$$C = \bar{z} - mx \, where \, \bar{z} = \frac{1}{N} \sum_{i=1}^{N} z_1 \qquad (10.4)$$

The process of estimating $m$ and $C$, the tangent and the intercept, can be regarded as an application of the principle of maximum likelihood assuming $N$ sets of random independent variables.

The calculation of the flatness and straightness deviations using formulae like (10.3) instead of (10.2) is becoming less appropriate as instruments become more integrated. Precision and miniature coordinate-measuring machines will soon be available to measure straightness etc. as well as size and there will be no guarantee that the general direction of measurement will be close to the plane or line direction being sought. This is a very important consideration for the future.

The problem with using least squares is that the numerical formula is not very stable. This is because the estimate of '$m$' for example or '$2\alpha$' depends on using the difference between two very large summations. This can be seen in Equations (10.2) and (10.3). It is recommended that double precision is used for the data points to avoid the '$m$' value being evaluated from numerical noise. A check on the calculation of '$m$' should use about the same number of data points as the actual experiment.

A very simple method of getting an estimate of the slope often used in practice is merely that of joining the first and last data points together with a line. Another method consists of defining a zone comprising two separated parallel lines, which just contain the straightness data between them.

A few points need to be noted. One is that such a zonal method needs three points of contact. It is easy to show that if there are only two points of contact, one at a peak and one at a valley, then the separation of parallel lines through them is not a minimum. Three points of constraint are needed, which could be two peaks and one valley or two valleys and one peak (Figure 10.14). It can be shown that in situations involving the minimization of linear problems, the maxima and minima will always be alternate. This demonstrates the principle of the Steifel exchange technique of linear programming. There are three unknown parameters, $m$, $c$ to establish one line and $E$, to establish the separation. Three contact points are needed (Figure 10.14 [10.2]).

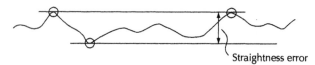

Straightness error

**Figure 10.14**  *Straightness – Steifel exchange*

How this zonal method differs from those used in roundness will be discussed in the relevant section on roundness. It is with regard to issues such as zones that Table 10.3 is useful [10.3].

Usually, but not necessarily, the zonal method described here will give a smaller actual value of the straightness than the least-squares method. Obviously, by definition it cannot ever be bigger because it is the minimum value. Differences between the two methods usually do not exceed 10% or thereabouts. Similar considerations apply to flatness. Also, the obvious problem with zonal methods is their susceptibility to the odd very large peak or valley. This is the same problem as that which occurred in the evaluation of surface texture peak parameters. The danger is that a large peak or valley may not be typical of the process.

It could be argued that alignment should be encompassed in the subject of straightness because it is the measurement of discrete points from a line. The basic technique is similar to that of straightness measurement but the art is in its application to a multiplicity of engineering components such as large frames, bearings and plant. These details are covered adequately elsewhere.

Errors in gears, screw threads, etc. are specialized subjects dealt with in engineering metrology books and will not be covered here. Needless to say the same problems always exist: that of establishing a reference and assessing the errors from it.

## 10.5  Flatness

Flat surfaces of large size such as surface plates and tables are of considerable importance in precision engineering because they act as reference surfaces for the inspection of other workpieces.

Assessment of the flatness of a surface can be accomplished in a variety of ways and with good fidelity. One of the oldest ways of testing a surface for flatness is to sweep it in several places with a straight edge and look for a chink of light. Diffraction of the light occurs through the gap it if is very small. Also, the edge has to be rotated on the surface to ensure true planarity (Figure 10.15).

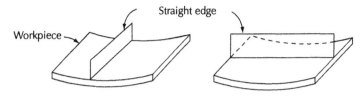

**Figure 10.15**   *Straightness – straight edge*

Other methods of measuring flatness such as the use of coordinating-measuring machines working relative to a fixed axis, or the use of autocollimators (Figure 10.7 (b)) to measure local angle levels and curvature devices, as previously used in straightness measurement, have their attendant problems of noise generation. For this reason, the data points have to be used with great care.

In addition to the technique discussed in the measurement of straightness, there are additional ones in flatness. The most important of these is interferometry [10.4], especially in the heterodyne mode, which enables the contour of the surface to be evaluated in absolute terms. This technique is described in the section on optical methods. It must suffice here to say that the technique is viable and does produce, under certain conditions, absolute values.

There are a number of ways of measuring flat surfaces. Amongst these are optical methods, including interferometry [10.4], holography [10.5] and moire [10.6].

Despite the potential of these methods, especially interferometry, they are not in general use in engineering because they always have constraints. Interferometers require optically smooth surfaces. In holography, the reference is generated by computer as a hologram. This reference, however, never has the same microstructure i.e. roughness as the real part, making the subtraction of wavefronts difficult.

Also, the test surface has to be bright and cannot be black as often are epoxy granites. Even cast iron presents problems.

Moire measurement requires projecting patterns onto the surface and viewing through a primary grating [10.2]. A breakdown of the various methods is given in [10.7].

One of the straightforward ways of getting a flat reference is to use liquid, usually water [10.8]. Problems with vibration and poor damping make the systems impractical. Mercury can be used but has health risks.

The idea of using a plane generated by rotating a line of sight has been used. One preferred method used in industry has a fixed alignment telescope with a rotating penta prism [10.9].

Flatness is an extension of straightness errors to two dimensions (called areal here, as in roughness).

One equation of a plane is:

$$z = c + m_1 x + m_2 y \qquad (10.5)$$

where $x$ and $y$ are here taken to be the causative variables and $z$ the observational variable.

There are a number of methods of assessing flatness, of which some are similar to straightness. However, because the assessment is somewhat laborious, there are often pilot measurements based on a 'Union Jack' pattern (Figure 10.16) that limit the measurements somewhat. Historically, it appears that the first reference plane was simply taken to be that plane defined by passing it through three corners of the plate. It has since been suggested that the plane parallel to the diagonals is becoming more widely used.

These two methods are not analytically defined and yet are relatively easy to carry out. The 'Union Jack' pattern is often criticized because of the relatively poor coverage of the area. However, as was seen in straightness, errors are often in the form of a bow and the maximum errors are generally at the edges of the plate or mirror where coverage is most extensive, so the choice of pattern is not so haphazard as might appear.

Not all flatness problems involve surface plates. More and more investigations are involved in the measurement of the flatness of rough objects. In these cases, autocollimators and inclinometers are not so much used. Probe methods can be used, which measure relative to a fixed internal datum plane or movement.

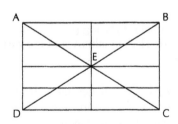

**Figure 10.16**   *Testing of engineer's surface plate*

Consider first the least-squares plane method. As an example, the technique adopted in straightness will be extended. The normal equations are similar to those in straightness but have one more variable in number. The variables of minimization are derived from making S – the sum of deviation squared – a minimum, that is,

$$S = \sum_{i=1}^{N} \left[ z_1 - (c + m_1 x_i + m_2 y_i) \right]^2 \qquad (10.6)$$

is a minimum.

The right-hand side is similar to that for straightness. From equation 10.5, $c$, $m$ and $mz$ can be found.

It can also be shown, as in straightness, that the best-fit plane always passes through the centroid of the observed points $x$, $y$, $z$. Referring all measurement to this as the origin therefore eliminates the term $c$ in the equations given (also true in straightness).

If uniform increments are taken in orthogonal directions, it is possible to simplify these equations considerably, yielding the new plane:

$$z = m_1 x + m_2 y \qquad (10.7)$$

where:

$$m_1 = \frac{\sum y^2 \sum xz - \sum xy \sum yz}{\sum x^2 \sum y^2 - \left(\sum xy\right)^2} \qquad (10.8)$$

$$m_2 = \frac{\sum x^2 \sum yz - \sum xy \sum xz}{\sum x^2 \sum y^2 - \left(\sum xy\right)^2} \qquad (10.9)$$

where the summation is over all points in the plane.

There are much simpler ways to establish a plane, however, and it has to be said that usually there is not much difference between estimates of flatness from any method. Perhaps the simplest is to join any three corners together with lines. The plane containing these lines is then used as the reference. This is called the **three-corner plane** and is often used in industry.

It has been pointed out that many of these sophisticated methods are unnecessary and uneconomic. For example, using three plates and hand scraping to remove high spots can give plates flat to about 5µm, so it may be only in cases where the surface plate suffers wear and needs to be checked that the more automatic inspection methods of testing are needed. How to achieve automatic inspection quickly is at present being investigated.

Another reason for using an optimized datum plane is that the manufacturers of plates could penalize themselves unnecessarily by reporting maximum deviations from a non-optimum plane. The actual flatness error value reported depends on the method of assessment used.

It is obviously worthwhile trying to get better methods of evaluation, even for this reason alone. The manufacturer then picks the plate that shows the smallest deviation, which then stands a better chance of being sold.

A great many algorithms have been devised in which the grid system takes into account the shape of the surface to be measured.

Figure 10.17 shows an example of triangular cells as opposed to rectangular ones as for the 'Union Jack' methods.

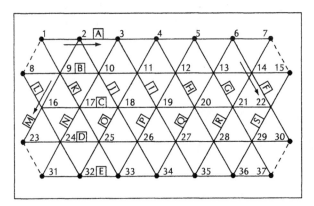

**Figure 10.17**   *Triangular measurement cell*

Ideally, the order in which the points are taken should be specified so as to reduce the errors – and to check on the build-up of errors e.g. in Figure 10.18.

Key positions for the data, particularly those on the diagonals, are taken with extra care, the verticals usually acting as checkpoints.

Another shape is shown in Figure 10.18. Notice that there is always some gap left at the edge of the plate. If wear has occurred on the plate, it is likely to be at the periphery and so it is never used for high-accuracy work.

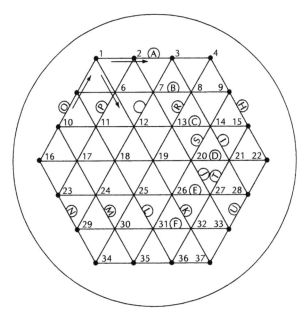

**Figure 10.18**   *Triangular cell to fit circular plate*

Triangular cells are being used more often because of the better coverage of the surface. Also, the possibility of hexagonal sampling patterns is now being tried because of good coverage and high accuracy.

## 10.6   Conclusions

Straightness and flatness are very important in engineering. Maudsley [10.10] regarded the flat to be the very base of engineering metrology. This helped Roberts to develop the 'planer' machine tool in 1817, which was one of the foundations of the Industrial Revolution in England.

Despite many advances, particularly in physical optics, the traditional methods using alignment and autocollimation are still used.

## References

10.1   Hume K. J. Sharp G. H. Practical Metrology Vol. 1 Macdonald London (1953).

10.2   Whitehouse D. J. *Handbook of Surface Metrology*. Inst Phy. Pub. Bristol (1994).

10.3   Chetwynd D. G. Ph. D Thesis Leicester (1978).

10.4   Korner K. *Messung der Ebenheitsabweichung mit einem Echtzeit* – Interferometer Feingertetechnik. Vol. 37 p15 (1988).

10.5   Vest C. M. *Holography Interferometry*. Wily New York (1979).

10.6   Reid G. T. *Moire Fringes in Metrology*. Optics and Lasers in Engineering Vol. 5 p63 (1984).

10.7   Meijer J. *From Straightness to Flatness*. Ph. D. Twente (1989).

10.8   Schultheiss Metaal bewerking Vol. 29 No. 2 p3–8 (1963).

10.9   Taylor Hobson Alignment Telescope with Accessories.

10.10  Hume K. J. *A history of Engineering Metrology*. MEP London (1980).

10.11  Whitehouse D. J. *Some Error Separation Methods in Surface Metrology*. Proc. Inst. Phys. J. Sci. Inst. Vol. 9 p531 (1976).

# 11

# Roundness and related subjects

## 11.1   General (See ISO 6318 1985 for terminology, 14291 for properties)

So far, deviations from ideal shapes, which are linear or substantially so, can be related to a straight generator. There are, however, many other different shapes that are equally important. In particular, the circle can be regarded as one such geometric element from which more complex engineering forms can be derived. About 70% of all engineering components have an axis of rotational symmetry in them somewhere. The importance of this was first realised by R. E. Reason of Taylor Hobson in 1951 [11.1], who set out almost single-handed to build instruments to measure roundness. Out-of-roundness or, more simply, roundness will now be considered as the next logical step in building up a more total geometry of a surface.

How the roundness error can be visualized with respect to surface texture errors is shown in Figure 11.1. Also shown in the figure are roughness marks C and waviness marks A and B.

Roundness or, strictly, out-of-roundness is more important than is usually thought. It is not generally realized that the energy needed to turn or grind comes from rotation; either from the workpiece or the tool. This rotation is absolutely critical to the process and the efficiency of the energy transfer is largely determined by the axis of rotation.

Control of the rotational axis is achieved by roundness measurement (Figure 11.2 (a)).

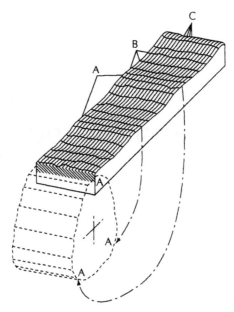

**Figure 11.1** *Roundness and texture*

Very many of the common machining processes are rotational e.g. grinding, turning and milling. Only planing and broaching are translational. Also, in functional situations, lateral movement between surfaces is usually of a rotational nature. Figure 11.2 (a) shows how roundness relates to manufacture and 11.2 (b) shows a journal bearing in which the roundness determines the fidelity of rotation.

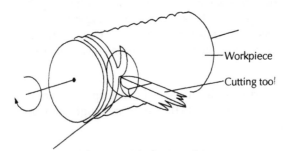

Workpiece

Cutting tool

Energy to cut material - from rotation

*Roundness provides control of rotation*

(Rotational energy used in turning, milling, grinding.
Translation energy only planing)

**Figure 11.2 (a)** *Manufacture and roundness*

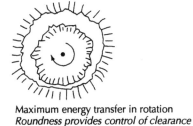

Maximum energy transfer in rotation
*Roundness provides control of clearance*

**Figure 11.2 (b)**   *Function and roundness*

Figure 11.2 (c) shows the function map described earlier. It shows how two basic surface metrology features, process and machine tool, link with roughness and roundness.

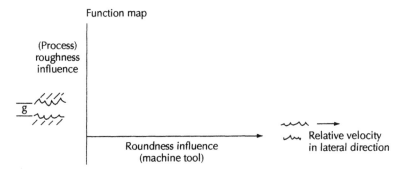

**Figure 11.2 (c)**   *The function map*

## 11.2   Direction of measurement

The roughness and roundness influence in the function map (Figure 11.2 (c)) are at right angles to each other. This is consistent with the direction of measurement being at right angles to each other on a round workpiece (Figure 11.3).

The process marks are at right angles to the path of the tool. Measuring with the 'lay' indicates machine tool effects and, in particular, the errors on the tool path. If the test piece is circular, this means measuring the roundness.

Also, it is important when measuring roundness to use a suitable stylus, otherwise roughness and roundness can get mixed up. In Figure 11.3, the hatchet stylus acts as a mechanical filter. It is of a deliberately long radius in the axial direction to ensure that the stylus does not drop into the process mark. Figure 11.4 shows the distortion in shape and position of the roundness graph when the sharp stylus penetrates the finish [11.1].

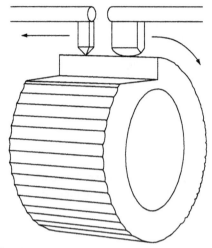

**Figure 11.3**   *Directions of measurement*

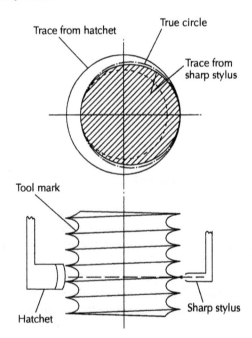

**Figure 11.4**   *Stylus for measurement*

## 11.3   Display of roundness

There are a number of ways of displaying roundness. One is Cartesian and another
is polar. Figures 11.5 and 11.6 show the differences. In general, the polar form is
used to get a better visual correlation between the object and the chart

representation. However, it is not such a straightforward choice as might be expected, as will be seen later.

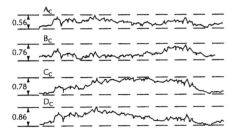

**Figure 11.5**   *Cartesian display of roundness*

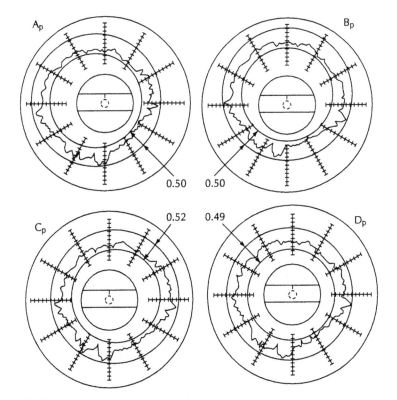

**Figure 11.6**   *Polar display of roundness*

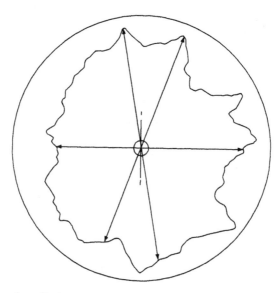

**Figure 11.7**   *Choice of coordinates*

On balance, the polar coordinates scheme has been preferred (Figure 11.7), provided that measurements are made through the centre.

The first problem is the definition of roundness. A multitude of alternatives are in use, some derived from the measuring instrument used and some from the use of the workpiece or its method of manufacture. Generally, a workpiece is described as round in a given cross section if all parts on the periphery are equidistant from a common centre (the Euclidean definition). This sounds obvious but in most cases the part has not got a diametrical form described completely by the equation of a circle. Superposed are other types of form. Some are due to the method of manufacture such as grinding marks and some are due to errors in the particular machine tool producing the part. These are the phenomena described earlier in surface texture. Usually, however, they are excluded in the primary measurement by means of filters or a blunt stylus acting as a mechanical filter. Out-of-roundness is usually restricted to a consideration of lobing and to some of the higher frequencies. It is in the identifying of the different harmonic components that the characterization takes place. In roundness, however, there is a need for the characterization of the method of collecting the data, as will be seen. Because of the many different sizes of workpiece it is usually best to specify frequency characteristics in terms of what is effectively a wave number, that is in terms of undulations per circumference of revolution (upr).

## 11.4   Lobing

Lobing is considered to be those deviations having a wave number from 2 to about 15, although sometimes higher numbers are included. These are often of quite high magnitude (e.g. micrometres in height) and most often of an odd wave number. These odd lobes are most frequently of almost constant diameter as a result of being made by centreless grinding. Lobing may exhibit an even or odd number of lobes, and may be of more or less constant height and spacing as produced by grinding, lapping, etc. but can be more random. Lobing is regarded as important, especially in the problems of fit. One basic consideration for fit is that the effective size variation for odd-lobed cylinders is positive for external profiles and vice versa for internal profiles. There is one big difference between odd and even lobing, and this is that even lobing is measurable by diametrical assessment but odd lobing is not. For some applications, this constant-diameter property is not important, for instance in ball bearings, where separation between components is the only criterion. Because one criterion used for out-of-roundness is the variation in diameter, a comparison between the properties illustrated in Figures 11.8 and 11.9 is informative. Figure 11.8 shows a constant radius and Figure 11.9 constant diameter workpiece. The latter is a Releaux triangle (spherical triangle). How this odd and even lobing works out will be considered in Section 11.5.2. Clearly Figures 11.8 and 11.9 are different yet both could be described as 'round'.

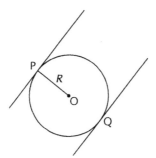

**Figure 11.8**   *Constant-diameter figure*

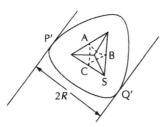

**Figure 11.9**   *Constant-diameter figure – Releaux triangle*

The properties of a Releaux triangle are significant. The generator PQ has two of the properties of diameters of circles. It has constant length and is normal at both ends to the tangent to the curve. The generator PQ constrains the true centre at only three angles and the radius from the centre 0 is only normal to the curve at six points. (In general, for an $n$-lobed figure there are $n$ and $2n$ points.) Notice also that the generator PQ is normal to the curve because the constraining curves AB, BC and CA are arcs of circles. One further point that can be significant in instrumentation is that the instantaneous centre of rotation of PQ is not the mid-point $R$ but one of the three apexes of the triangle, for example point S as shown in the figure.

**Note**: the Releaux triangle has the smallest area and the greatest deviation from roundness of all constant-diameter figures.

Higher-order lobing of up to nearly one thousand undulations per revolution is said to be important from a functional point of view, especially again in the bearing industry where acoustic noise may be important.

Another common form of out-of-roundness is ovality or the two-lobed effect. Multiplicity of this effect is sometimes called **polygonation**. In some circumstances, parts may need to be oval, for instance in piston rings, so that out-of-roundness errors may not necessarily be detrimental to performance. Oval or elliptical parts are unique in roundness because they can easily be made circular simply by the application of a force of compression on the long diameter, or vice versa. Other names have been given to different features of the spectrum of undulations. One used when there has been chatter established between the grinding wheel and workpiece is called **humming** and owes its name to the acoustic noise that it produces and results in a characteristic peak in the spectrum of the workpiece.

Roundness means different things to different people. This is why the characterization is important. There are three basic ways in which roundness can be measured.

## 11.5   Methods of measuring roundness

### 11.5.1   Fourier analysis

Fourier analysis is one of the principal analytical tools of metrology. It has been mentioned earlier on in random processes when characterizing the roughness and waviness of a workpiece. One aspect of the analysis are the Fourier series, which are used for characterizing periodic waveforms. Roundness is a perfect example of a periodic wave. The objective of the analysis is to find the presence and value of lobes. Fourier series analysis can do this. First, the geometric signal is obtained, then from this, by an operation, the number of specific lobes can be determined along with their relative phases (position) within the period. Fourier analysis can be used to help understand roundness measurement.

Thus, Figure 11.10 shows a shape between two anvils corresponding to, say, a micrometer. The distance between the anvils is $r_1 + r_2$ or, more correctly, $r_1(\theta) + r_2(\theta - \pi)$ (see Figure 11.13), where $(\theta - \pi)$ is the position of $r_2$ relative to $r_1$. In this case, it is $-\pi$ out of phase.

This obvious equation can be put in Fourier terms $F_1(w)$ and $F_2(w)$. In the case above, the amplitude of $r_1$ is the same as $r_2$ but with a phase shift of $\pi$. This value is $F_n(w)$ – the amplitude of $r_1$ and $r_2$ – and a phase shift in Fourier terms of $\exp(jn\pi)$ where $n$ is the number of lobes per revolution of interest.

Putting the geometric situation into the Fourier form makes it easy to see possible effects of the measuring equipment. Thus, for $n = 1$, $\exp(j\pi)$ is the phase that is $-1$ so the Fourier coefficient for $r$ is:

$$F(w) \text{ and } r_2 \text{ is } F(w)exp(j\pi) = F(w)(-1) \qquad (11.1)$$

So, $r_1 + r_2$ has a Fourier component for $n = 1$ of $F(w)(1-1) = 0$. For $n = 2$, an ellipse-like figure, the coefficient is $F(w)(1 + \exp(j2\pi)) = 2F(w)$ – the Fourier coefficient does exist so that out-of-roundness of ellipse or oval character will be seen by the diametrical method. For $n = 3$, the coefficient is again zero.

It works out that for parts with even lobing, the part can be measured with a micrometer type instrument but if the part has odd lobes it cannot be measured with a micrometer.

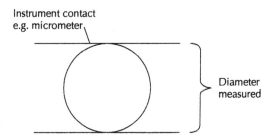

Figure 11.10   *Diametrical methods*

As indicated, the micrometer type of measuring instrument is only suitable for measuring parts having an even number of lobes per revolution. Consider an ellipse or oval part as in Figure 11.11:

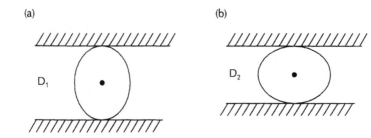

**Figure 11.11**    *Diameter of even-lobed figure*

As the part is rotated, the reading varies from $D_1$ to $D_2$. So, the diametrical variation of the part is indefinite and varies with position.

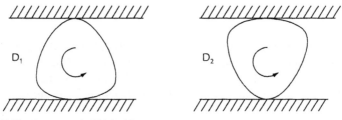

**Figure 11.12**    *Diameter of odd-lobed figure*

For a three-lobed figure, the diameter is constant and by using this instrument the part looks circular (Figure 11.12).

Bodies needing constant diameter i.e. rollers in roller bearings have to be odd-lobed. Even more obvious is the fact that coins to be used in vending machines have to be odd-lobed otherwise they run the risk of jamming in the slot.

### 11.5.2  Diametrical methods

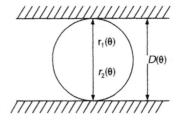

**Figure 11.13**    *Radial representation*

Summarizing, when $n$, the number of undulations per revolution, is an odd number, then $F_m(n)$ is zero – despite the fact that $F_{actual}(n)$ exists. Hence, diametrical methods cannot see bodies with an odd number of lobes. As mentioned before, this might not be serious if the body is to be used as a spacer such as in a roller bearing, but it would be serious in, say, a gyroscope, which spins about its centre (Figure 11.13).

### 11.5.3   Chordal methods

The second method represents the traditional vee-block method (Figure 11.14). Strictly, it is the variation in the chord joining the points of contact of the part and the two faces of the vee that is measured when the part is rotated. This variation induces changes in the vertical position of the gauge when the part is rotated.

Note that chordal methods can only be converted into radial results from a common centre when undulation spacing is uniform and the vee angle is chosen correctly.

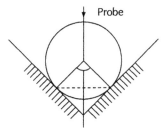

**Figure 11.14**   *Vee-block measurement of roundness*

As the part is rotated, the indicator reading varies. The difference between the maximum and minimum readings is noted. This is called the **total indicator reading (TIR)**.

It is interesting to note that this technique is a derivative of the intrinsic equation method of defining an arbitrary curve relative to itself by means of a tangent at P and the length of curve from an arbitrary axis. Here, the tangent is where the vee contacts the part and the arc length is the circumferential distance between the two points of contact [11.2].

As in the case of the diametrical method, this technique introduces distortions into the assessment of roundness. These need to be understood because of the general use and misuse of the vee-block approach in industrial workshops. The distortion of the lobes in this case is given by:

$$F_{measured}(n) = F_{actual}(n)\left(1 + (-1)^n \, \frac{\cos n\gamma}{\cos \gamma}\right) \tag{11.2}$$

For example, if $n = 5$ and $\gamma = 60$, then $F(5)$ gives a measured value of zero although it is obvious that the workpiece is not perfectly round! Equation (11.2) is again a Fourier equivalent of the geometric signal. Clearly, using any vee method introduces some distortion but this can be reduced if more than one vee is used. Where one vee has distortions, another can be used to give a more realistic value. Two vees have been used as a commercial instrument but found only a limited market [11.3].

There are other methods of chordal roundness. Figure 11.15 shows a German method in which an extra degree of adjustment is obtained by having an offset probe at angle $\beta$.

From calculation, it was found possible to choose an offset angle and a vee angle that had a minimum envelope of errors over lobes of $3 \rightarrow 15$ upr. The best vee works out at about $110°$ and the offset angle at $6°$.

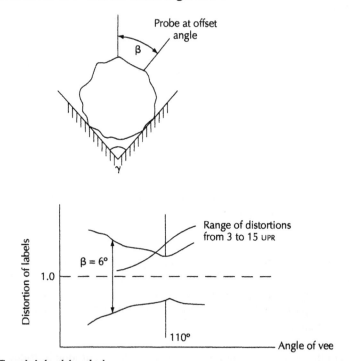

**Figure 11.15** *Extended chordal method*

There are various other alternatives shown in Figures 11.16, 11.17 and 11.18.

All of these methods have been used industrially. The multiple floating skid method (Figure 11.16) is used in Russia and Finland for measuring rolls for paper mills. They are, in effect, generating intrinsic references by using multiple skids.

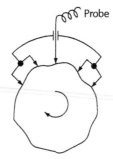

**Figure 11.16**  *Multiple floating skid method*

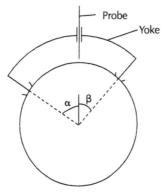

**Figure 11.17**  *Three-point roundness with two skids*

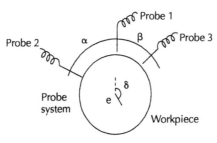

**Figure 11.18**  *Variable error removal in roundness*

Assessment of roundness using two- or three-point contact are considered in ISO standard 4292:1996. The more the number of skids, the better the reference. However, too many skids makes the apparatus cumbersome and mechanically unstable.

Figure 11.18 shows a method in which each of the probes are active i.e. none are skids. Furthermore, each probe has a different gain to the others and the angles between them are unequal. It is therefore possible to arrange gains and angles such that when the signals are added together, the sum of the signals will not 'see' any movement of a perfect circle touching the probes. In other words, the method does not require a good axis of rotation. The penalty for allowing a poor axis of rotation is that the true out-of-roundness signal has to be computed out of the sum signal. All the other harmonics will also be distorted but the amount of distortion is known and the coefficient compensated. When this has been done to all the coefficients of lobing in the spectrum, the compensated coefficients are added up as a Fourier series and then retransformed to give the out-of-roundness of the workpiece. The only proviso is that the axis of rotation is continuous and preferably uniform.

Another measurement using the chordal method is shown in Figure 11.19 [11.4]. This has one vee angle but two probes as shown. The two probes are connected differentially to measure $dr/d\theta$ – the radial rate of change. This is used by some ball and roller bearing manufacturers for assessing acoustic noise rather than using a roundness value.

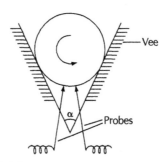

**Figure 11.19**   *Measurement of radial slope*

### 11.5.4   Radial methods

Changes in radius relative to a true circle are what is usually regarded as roundness error. So, instruments measuring out-of-roundness have to provide a circle somewhere, whether in hardware or software to act as the reference. The following types of instrument are in general use.

Figure 11.20 (c) is a comparison method that does not rely on a precision spindle as do (a) and (b). Instead, the reference is a master disc clamped onto the test workpiece. As the workpiece is rotated, the difference signal from the two transducers is taken as the error in roundness. It does not matter if the rotation axis is poor at either end, providing that the master disc and the workpiece with their probes are close together. Errors due to Abbé separation are then second order.

The following truly radial methods were invented and developed almost single handed by R. E. Reason of Taylor Hobson in 1951. They represent the first successful roundness measuring instruments.

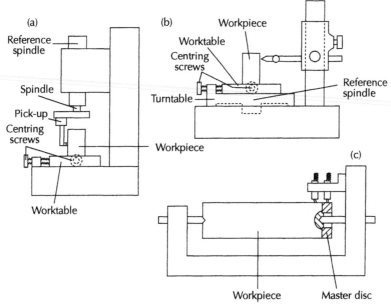

**Figure 11.20** *Radial methods*

Illustrated in Figure 11.20 (a) is the **rotating spindle method**. In this, the transducer is mounted onto the precision spindle above the workpiece. The idea is that the one arm of the transducer acts as the reference arm of the calliper, which is linked directly to the spindle. This generates a perfect circle in space (as good as the spindle). The other arm of the calliper contacts the workpiece. The difference in movement between the two arms as the spindle rotates is sensed and magnified and represents the out of roundness of the workpiece (Plate X (a)). Plate X (b) shows close up views of the instrument. Plate X (b) shows the rotating spindle type of instrument close up, including centring and levelling table. Plate XI shows a portable rotating spindle instrument that is carried to the workpiece. Plate XII shows a rotating table type of instrument – for large workpieces.

The other alternative is the rotating table where the precision spindle supports the table on which the specimen rests. In this, any point on the table generates the reference arm of the calliper. The specimen uses this perfect rotation to contact the probe through 360°. Errors in the specimen spoil the circular accuracy generated by the table. The rotary table range available for general use is very large.

Which is the best configuration for measuring roundness: the rotating spindle or the rotating table? The answer depends on the application.

The main difference between the two methods is that eccentricity, concentricity and squareness are more easily measured by the rotating table method and radial variations with height (i.e. the size) are best measured with the spindle table.

These will be described below but will also be discussed in considering the measurement of cylindricity in Chapter 12.

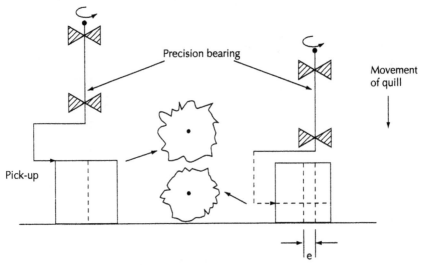

**Figure 11.21** *Rotating spindle – change in size*

Figure 11.21 illustrates the fact that despite moving the pick-up vertically to measure different locations, the measurement of the change in radius from one level to the other is valid. What is not valid is any measure of eccentricity. Fidelity of part centre is lost if either the pick-up or the part is moved relative to each other.

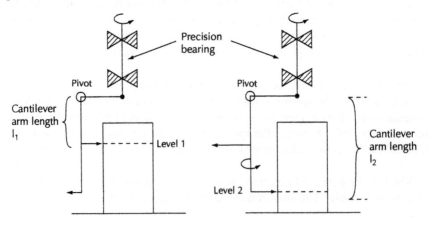

**Figure 11.22** *Stylus tree*

There is one way of maintaining spatial fidelity using the rotating pick-up method. This is shown in Figure 11.22. Two (or more) levels are measured by styli that are positioned at the correct position on the same stylus stem. To measure different levels, the stem is twisted around until the appropriate pick-up makes contact (Figure 11.22 (b)). In this way, the position of the precision bearing is not changed relative to the part so that eccentricities and concentricities can be measured. The only problem is that the magnification of roundness seen at position 1 is different from that at position 2 by $l_1/l_2$, which has to be corrected at a later stage.

Size variation with height cannot be measured easily with the rotating table method because if there is a squareness error between the axis of the spindle and the column supporting the probe, this can give an apparent size change of the part when the position of the probe is moved, as shown in Figure 11.23.

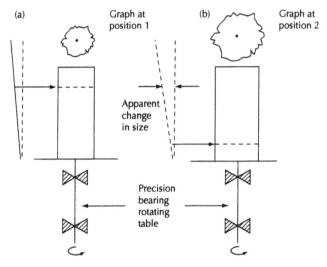

**Figure 11.23**   *Rotating table – size problem – importance of squareness*

It is possible to get rid of squareness errors by taking two measurements at each position: one is the usual trace at position 1. The other is at position 2, where the pick-up is moved through the centre of rotation and reversed in direction of sensitivity. The true trace is simply (trace 1 + trace 2) divided by 2 (Figure 11.24).

Eccentricity and related features are preserved using the rotating table method because the part is not moved relative to the precision bearing when moving from one level to another. Also, the column should be straight, otherwise error in the level and direction of sensitivity can result (Figure 11.25). Taking these two problems into account, in addition to the fact that the accuracy of the bearing depends on the weight and position of the part as shown in Figure 11.26, makes it difficult

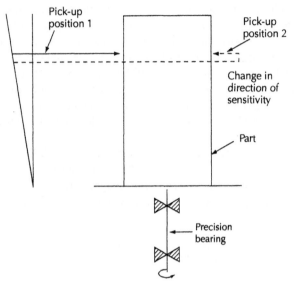

**Figure 11.24**   *Rotating table – size problem – importance of squareness*

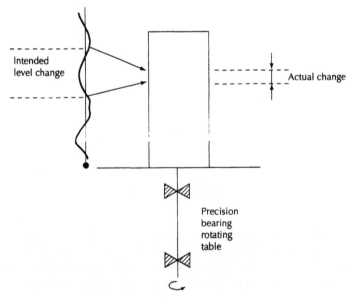

**Figure 11.25**   *Rotating table – importance of straightness*

to understand why this configuration is often preferred to the rotating pick-up method.

The reason for the popularity of the rotating table has to be the ability to measure eccentricities, concentricity and squareness easily. These features are most often required when measuring cylindricity, hence the wide variety of instruments that are needed to satisfy customer requirements. Some differences in specification and dimension are given in Plates XIV, XV and XVI. What is sometimes forgotten with the rotating table method is that in the simpler instruments, centring, and levelling can be a problem because the table with the adjustment is rotating. It is, however, straightforward to make the adjustments automatically.

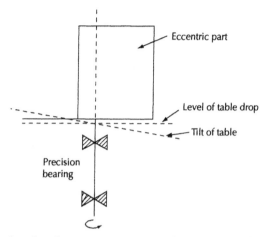

**Figure 11.26**   *Effect of weight and eccentricity on accuracy of rotating table method*

### 11.5.5   *Metrology schemes of roundness instruments*

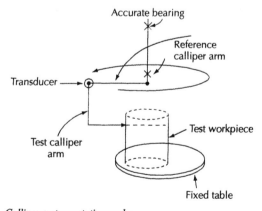

**Figure 11.27 (a)**   *Calliper system rotating probe*

It is quite difficult sometimes to see the basic elements in a metrology instrument and roundness measurement is no exception. Figures 11.27 (a) and 11.27 (b) (overleaf) show the essential calliper arrangement. It is clear that in the rotating

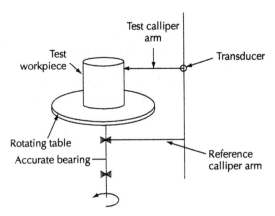

**Figure 11.27 (b)** *Calliper system, rotating table*

spindle configuration, the actual rotation mechanism is separate from the workpiece, whereas the workpiece is actually resting on it with the rotating table configuration. This has a profound effect on flexibility of use, which is shown in Plate IX. This shows the case where the rotating spindle type of instrument is taken to the work. The figure shows the quill (the housing of the precision bearing) inserted into a cylinder. Roundness measurements inside the cylinder at different levels are therefore easily possible. The reason for this flexibility is perhaps best seen diagrammatically in Figure 11.28. Figure 11.28 shows the metrology diagram for the rotating table – being the simpler version of Figure 11.27 (a). This is the same figure as 11.27 (b) but for a different reason.

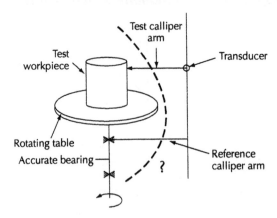

**Figure 11.28** *Rotating table – the test and reference arms cannot be separated*

## 11.6   Nature of the roundness signal

### 11.6.1   Magnification and zero suppression effects

Because the roundness instrument mechanically generates a true circle in space, all signals when magnified refer back to this circle and not to the centre of the part. The result of this is that some simple shapes, such as a square, become distorted in a nonsensical way. Apparently, concave shapes can result from magnifying convex ones. This is absolutely valid and reflects the fact that it is the outer 'skin' of the part that is being magnified and not all the part. Restoring all the part gives back the reasonable shape.

It has to be stated here that although roundness instruments produce some strange effects, there is no alternative to their use. This is because the size of roundness marks is so very small when compared with the dimensional size of the typical part; magnifying everything to the value where the roundness errors are visible would produce an enormous chart.

Consider Figure 11.29. It shows a true circle encompassing a workpiece that is not quite round. As only the deviation from the circle is magnified as the magnification goes through 2×, 4× and 8×, the workpiece apparently changes shape from being convex to concave. This is absolutely genuine and occurs in all roundness instruments. If the probe magnified from the centre of rotation to the outside 'skin' of the surface where the out-of-roundness is, the chart size would be enormous as indicated in Figure 11.30. Notice also that a perfectly sensible ellipse becomes like a dumbbell when measured with a roundness instrument (Figure 11.31).

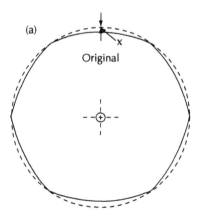

(a)

x

Original

**Figure 11.29**   *Effect of magnification – the surface skin only is magnified*

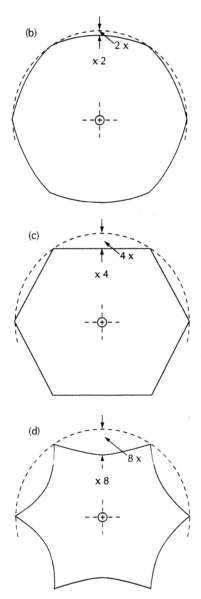

**Figure 11.29 (continued)** *Effect of magnification – the surface skin only is magnified*

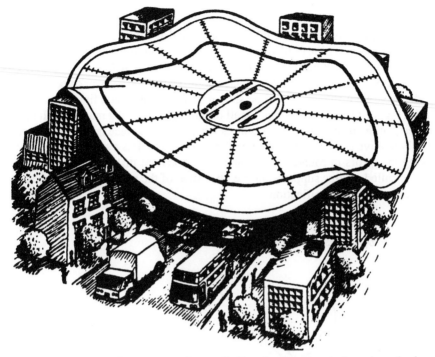

**Figure 11.30**   *Reason complete part cannot be magnified in order to see the out-of-roundness clearly*

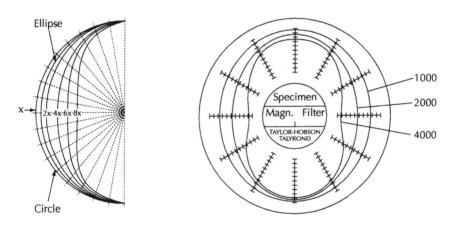

**Figure 11.31**   *Shape change of ellipse when magnified*

## 11.6.2   Representation by the limacon

Many of the problems surrounding roundness measurement are due to misunderstanding of the nature of the measurement. It seems natural to expect that if a perfectly circular part is measured, it should look like a circle on a graph. For the special case of roundness assessment, this is not so because it is not the whole part, say a shaft, which is being measured and amplified as would be the case if a coordinate measuring machine (CMM) is used for the measurement. A roundness instrument measures the peripheral 'skin' of the workpiece, not the whole part (see Figure 11.32). It is the disproportional magnification of the workpiece size that produces strange shapes. It will be shown that this shape is a limacon and not a circle (Figure 11.33). This is true for all roundness measuring machines! It has the property of distorting angles and centres when the workpiece is positioned eccentrically relative to the axis of rotation of the instrument. It is a small price to pay for having the errors in roundness revealed. The apparent distortions are a result of what is called 'radius suppression' i.e. the 'size' of the part is not magnified, it is removed prior to magnification.

### 11.6.3  Properties of the limacon

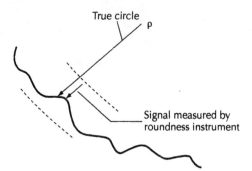

**Figure 11.32**  *Measurement of the peripheral 'skin' of the workpiece*

The true situation is shown in Figure 11.33, which represents a circular part that is eccentric.

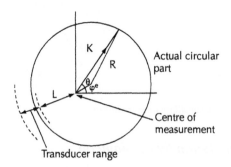

**Figure 11.33**  *Effect of eccentric positioning of workpiece*

Spatial situation: $k = e\cos(\theta - \varphi) + \sqrt{R^2 - e^2 \sin^2(\theta - \varphi)}\,(circle)$    (11.3)

This is the equation of an off-centre circle.

$$k(\theta) = e\cos(\theta - \varphi) + \left[R^2 \sin(\theta - \varphi)\right]^{\frac{1}{2}}\ldots$$

$$k(\theta) = e\cos(\theta - \varphi) + R - \frac{e^2}{2R}\sin^2(\theta - \varphi) +\ldots$$

$$\rho(\theta) = M(k(e) - L + S\ldots\text{ This is what the instrument sees.}$$

$$\rho(\theta) = M[e\cos(\theta - \varphi) + M(R - L) + S] - \frac{Me^2}{2R}\sin^2(\theta - \varphi) +\ldots$$

$$\rho(\theta) = M[e\cos(\theta - \varphi) + M(R - L) + S] - \frac{M^2}{2R}\sin^2(\theta - \varphi) +\ldots$$

$$\rho(\theta) = M(R - L) + S + Me\cos(\theta - \varphi)\text{ and letting } t = M(R - L) + S$$

$$\rho(\theta) = t + E\cos(\theta - \varphi)$$    (11.4)

This is the limacon shape seen by the instrument. The graphical equivalent of this effect is shown in Figure 11.35, where (a) is the real situation and (b) is what the instrument works on.

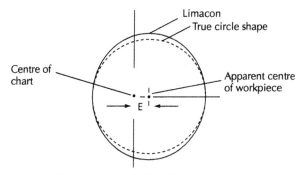

**Figure 11.34**  *True shape of a limacon compared with a circle*

Figure 11.34 shows the actual shape of a limacon when compared with a circle. It appears to bulge at right angles to the direction of the eccentricity.

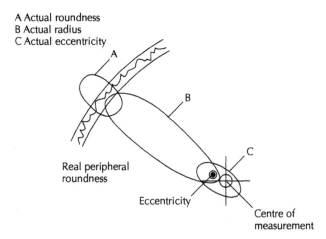

A Actual roundness
B Actual radius
C Actual eccentricity

A

B

C

Real peripheral
roundness

Eccentricity

Centre of
measurement

**Figure 11.35 (a)**   *Real situation*

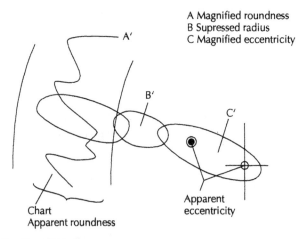

A Magnified roundness
B Supressed radius
C Magnified eccentricity

A′

B′

C′

Apparent
eccentricity

Chart
Apparent roundness

**Figure 11.35 (b)**   *Virtual situation*

All roundness instruments change the ratio of A:B:C in order to make A visible.

It should be pointed out that when the part is centred, the limacon and the circle are the same shape. In some respects, the limacon curve is easier to deal with than that of a circle. The circle is a quadratic term i.e. it has squared variables in the formula. The limacon is a linearized approximation to it – produced by the mode of operation of the instrument.

An interesting point is that the eccentricity term in the formula is particularly simple in the case of a limacon. It is the first term of the Fourier series representing the curve.

So, working out the first harmonic term in a roundness profile as found by a roundness measuring instrument automatically gives the degree to which the part has to be adjusted to be centred.

Equations (11.3) and (11.4) show how the components of the eccentric circle differ from the terms in the limacon.

Some people find it difficult to accept that if a perfectly circular part is not centred on a roundness instrument it should not look 'circular'. It looks like a limacon shape that bulges relative to a circle in a plane perpendicular to the direction of the eccentricity (see Figure 11.34). It has been asserted that by using a roundness machine, an eccentricity even on a perfectly circular part will make its shape appear to be a limacon. This is definitely not circular. It is possible in a computer to recalculate the components of the limacon, knowing the value of the eccentricity and rough size of the part to make the limacon look *circular and off-centre* as shown in Figure 11.36.

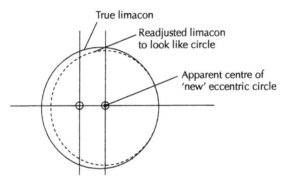

**Figure 11.36** *Readjustment of limacon*

It can even look reasonable to put a centre to this 'new' circle. This step is false – the centre of the real part has not moved, so all angles and measurements of diameter still must go through the chart centre (Figures 11.35 and 11.36).

If the software is arranged to 're-centre' the 'new' circle, then its centre will coincide with the true centre and there is no violation.

Angles have to be measured through the chart centre in all cases except when the suppressed radius term is restored to the signal and magnified. It is far better just to leave the limacon shape representing the part and accept that it is not circular.

There are other factors that present themselves. One is that the peaks and valleys on a roundness profile always point towards the centre of the chart and not the centre of the profile. This is seen in Figure 11.37. Slope measurement is also affected by angular distortion (Figure 11.38).

Perhaps the only case where a limacon (modified to a circle) is useful is if the roundness value is being measured off the chart by means of compasses and rule. It is perhaps just as easy to use a limacon compass and measure on the normal graph.

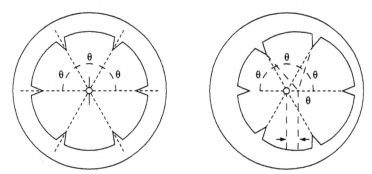

**Figure 11.37**   *Effect of angular distortion*

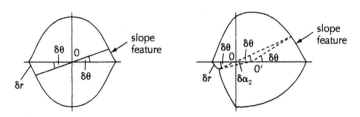

**Figure 11.38**   *Effect of angular distortion on slope measurement*

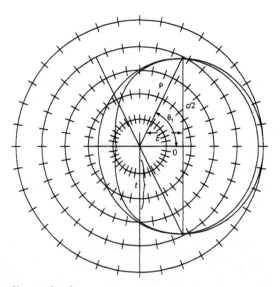

**Figure 11.39**   *Use of large polar chart*

The effect of the limacon shape is also dependent on the chart dimensions. If the chart centre is very small and the range of the transducer is very large i.e. the chart can be large as shown in Figure 11.39, then larger eccentricities of the workpiece relative to the axis of rotation can be allowed and the distortion shows up more. Keeping the transducer range small keeps the centring within small bounds but is inconvenient in setting up.

## 11.7 Assessment of roundness

### 11.7.1 The reference

Some useful ISO references for roundness are:

ISO 6318:1985    Measurement of roundness – terms definitions and parameters.

ISO 4291:1985    Methods for the assessment form roundness – measurement of variations in radius.

ISO 4292:1985    Methods for the assessment from roundness – two- and three-point method.

There are four ways of assessing the roundness:

(a) least squares;

(b) ring gauge;

(c) plug gauge;

(d) minimum zone.

The **least-squares method** is the most straightforward in the sense that all the data of the profile is used to establish the centre. The radius of the best-fit least-squares circle is found also from all the data. Actually, it is the best fit limacon that is measured. The largest peak to valley deviation from this 'circle' is then computed. This method is unique and therefore to be preferred. The only criticism is that it is difficult to measure the roundness error of a part from the chart using this method. Today, the use of the chart is minimal but before the availability of computers this was a serious drawback. Some problems with least squares will be shown later.

**Ring gauge** (or minimum circumscribing circle). In this, a circle is drawn around all the data. It is then shrunk onto the data to get a minimum size circle. This has to have just three points of contact between the data and the circle. The roundness error is simply the largest valley measured from this circle. This method is also unique but it suffers from the fact that the centre position is determined by only three points. So, a large untypical peak can produce an erroneous centre position, which can invalidate eccentricity and concentricity measurements.

**Plug gauge** (or maximum inscribed circle). This is the opposite of the ring gauge method. A 'circle' is drawn within all the data and is expanded until it is constrained by three valleys. The out-of-roundness is then the largest peak measured from this circle. Unfortunately, the centre obtained by this method need not be unique. A good example of ambiguity would occur if the plug gauge method were to be applied to the inside of a dumbbell shape. There are two equally good centres in this case (Figure 11.40).

**Minimum zone method**. In this, concentric circles are drawn, one outside the data and the other inside. These circles are adjusted until the 'zone' – the difference in radius of the two circles – is a minimum. In this circumstance, there are two points of contact on the outer circle and two on the inner – alternately contacting. This method has some ambiguity brought about by the inner circle definition in the same way as the plug gauge method but it is not usually considered to be a severe problem. The actual 'zone' is not ambiguous. It is simply the smallest in value. This smallest zone is the roundness error. These are shown in Figures 11.41 (a), (b), (c) and (d).

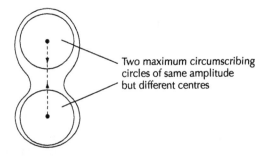

Two maximum circumscribing circles of same amplitude but different centres

**Figure 11.40**   *Problem with plug gauge method*

In general, the minimum zone method is quite difficult to find graphically and computationally. It has been criticized as being a hybrid method. The least-squares circle and centre correspond to the position found by a spinning shaft and the two gauge methods mentioned above relate closely to tolerance evaluation.

It is important to realize that none of these methods are right or wrong. As long as there is consistency in application between the maker and the user of parts, there should be little difficulty.

So far as the actual value of roundness is concerned, the minimum zone method by definition is the smallest. The least-squares value is about 5% higher on average and the two gauge methods 5% higher still.

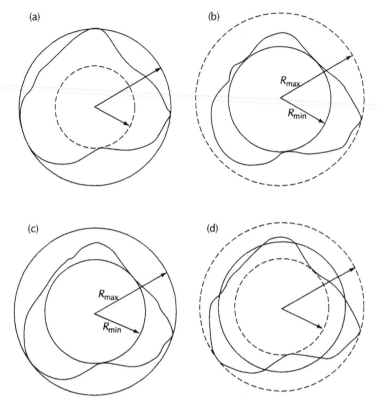

**Figure 11.41**   *Assessment of roundness (a) ring gauge, (b) plug gauge, (c) minimum zone (d) least squares*

The big difference lies in their variability. The least-squares method gives the most reliable roundness values because all the data is used (see Figure 11.42). The two gauge methods are the least reliable because of their dependence on the 'extremes' in the data. The minimum zone method lies in between.

Of these four methods of assessment, the least-squares method is easiest to determine.

Summarizing, the only way to get stable results for the peak-valley roundness parameters is by axial averaging i.e. taking more than one trace. The least-squares method gets its stability by radial averaging i.e. from one trace.

The calculation of the least-squares parameters is straightforward. Values of the signal are taken at equal angle increments through the 360° (see Figures 11.43 and 11.44).

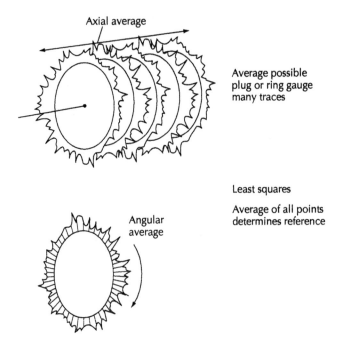

Axial average

Average possible
plug or ring gauge
many traces

Least squares

Average of all points
determines reference

Angular
average

**Figure 11.42**   *Reliability of circular reference systems*

## 11.7.2   Assessment of best fit parameters

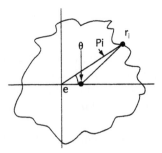

**Figure 11.43**   *Calculation of the least-squares parameters*

The radius term $R = \dfrac{1}{N} \sum_{i=1}^{N} r_i$    (11.5)

Where the $r$ values are taken from the chart centre around 360° back to the starting point, there are $N$ values in all. They are usually, but not necessarily, equal angles.

The coordinates of the centre are $\bar{x}$ and $\bar{y}$ where:

$$\bar{x} = \frac{2}{N} \sum_{i=1}^{N} r_i \cos \theta_i$$

$$\bar{y} = \frac{2}{N} \sum_{i=1}^{N} r_i \sin \theta_i$$

(11.6)

This is unique and the maximum peak and valley from it can be derived with confidence.

The minimum zone method is unique so far as the value of the zone is concerned but the centre position need not be unique as a result of the plug gauge problem. The problem is illustrated in Figure 11.45. Note that in all these procedures for roundness parameters, it is the centre that is often the most important parameter to determine and not the out-of-roundness value.

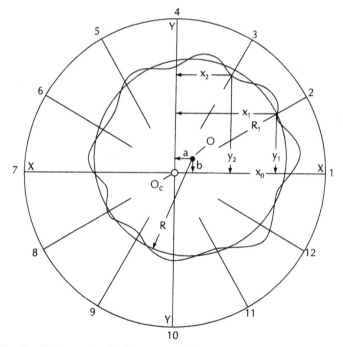

**Figure 11.44**   *Graphical procedure for determining the least-squares centre and circle*

There is a problem sometimes if the assessment of the roundness is measured on a graph. A slightly elliptical part if over-magnified as in Figure 11.43 will produce the dumbbell shape and consequently give two centres rather than the one if the plug gauge (maximum inscribed circle) is used. This is totally unacceptable.

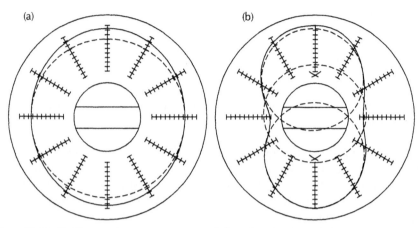

**Figure 11.45**   *Problems in plug gauge assessment of ellipse (see also Figure 11.31)*

It has been argued that the minimum zone method is unsatisfactory, being in between the ring gauge and plug gauge methods. Figure 11.46 tends to dispel this argument. Take the case of a bearing. The separation of the shaft from the journal is the critical parameter. So, if the clearance is being based on the way in which the roundness assessment of the shaft interacts with the roundness assessment of the journal, the various methods have quite different constraints. Consider the obvious case of ring gauge assessment for the shaft together with plug gauge for the journal. The interaction is determined by the interaction of the large valleys from the ring gauge with the large peaks from the plug gauge. In the minimum zone method, shaft and journal are assessed by minimum zone. Separation is where the two zones meet. It can be seen from the diagram that the interaction of the zones is twice as well established as the peak and valley extreme values of the gauge method. There are four points available to define the interaction with minimum zone but only two with the gauges.

Some of the next figures demonstrate the fragility of roundness assessment using peaks and valleys.

**Note 1:**

(1)  The energy interaction takes place at the common zone.

(2)  Common zone plug/ring: 2-point interaction.

(3)  Common zone minimum zone: 4-point interaction.

Figures 11.47 and 11.48 show how peak and valley positions and freak values are important.

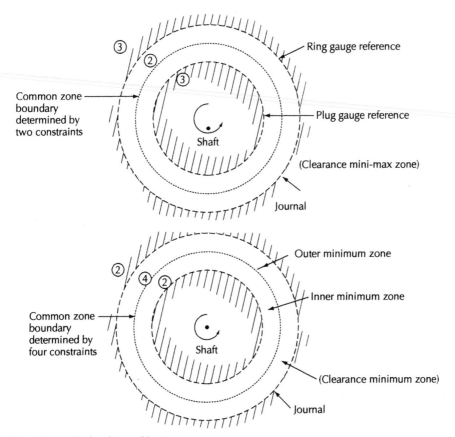

**Figure 11.46**   *Shaft and journal bearing*

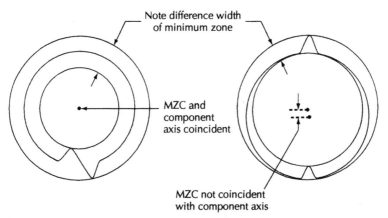

**Figure 11.47**   *Effect of relative positions of peaks and valleys on zone width*

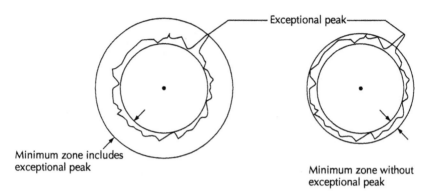

**Figure 11.48**   *How a single large peak can increase the apparent out-of-roundness*

## 11.8   Partial arc determination [11.5]

There are some critical cases in which there is not a full circle to assess. One example of this is when there is a keyway in a shaft as shown in Figure 11.50 (b). An even worse situation is when only a small arc is available as in Figure 11.50 (a). This small arc occurs in split bearings where a gothic arch is formed (Figure 11.49). Here, $r_1$ and $e$ are required.

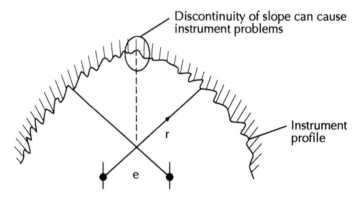

**Figure 11.49**   *Gothic arch*

It is not very convincing to derive the partial circumference parameters from this limited amount of data. A case for using a minimum zone could be made, perhaps, for the keyway situation but not for the gothic arch. In these circumstances, the least-squares method should be used. The calculations for radius terms and centre positions are tedious but tractable as seen in equations (11.7), (11.8) and (11.9).

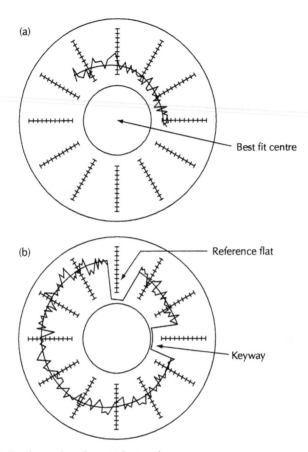

**Figure 11.50**   *Best-fit reference lines for partial circumference*

Thus:

$$\bar{x} = A\left(\int_{\theta_1}^{\theta_2} r\cos\theta d\theta - B\int_{\theta_1}^{\theta_2} rd\theta\right) + C\left(\int_{\theta_1}^{\theta_2} r\sin\theta d\theta - D\int_{\theta_1}^{\theta_2} rd\theta\right)\Bigg/E$$

$$d\bar{y} = \left[F\left(\int_{\theta_1}^{\theta_2} r\sin\theta d\theta - D\int_{\theta_1}^{\theta_2} r\,d\theta\right) + C\left(\int_{\theta_1}^{\theta_2} r\cos\theta d\theta - B\int_{\theta_1}^{\theta_2} r\,d\theta\right)\right]\Bigg/E \qquad (11.7)$$

$$\bar{R} = \frac{1}{\theta_2 - \theta_1}\int_{\theta_1}^{\theta_2} r\,d\theta - \frac{\bar{x}}{\theta_2\theta_1}\int_{\theta_1}^{\theta_2}\cos\theta\,d\theta - \frac{\bar{y}}{\theta_2\theta_1}\int_{\theta_1}^{\theta_2}\sin\theta\,d\theta$$

In this equation, the constants A, B, C, D, E and F are as follows:

$$A = \int_{\theta_1}^{\theta_2} \sin_2 \theta \, d\theta - \frac{1}{\theta_2} - \theta_1 \left( \int_{\theta_1}^{\theta_2} \sin \theta \, d\theta \right)^2$$

$$B = \frac{1}{\theta_2 - \theta_1} (\sin \theta_2 - \sin \theta_1)$$

$$C = \frac{1}{\theta_2 - \theta_1} \int_{\theta_1}^{\theta_2} \cos \theta \, d\theta \int_{\theta_1}^{\theta_2} \sin \theta \, d\theta - \int_{\theta_1}^{\theta_2} \sin \theta \cos \theta \, d\theta$$

$$D = \frac{1}{\theta_2 - \theta_1} (\cos \theta_1 - \cos \theta_2)$$

(11.8)

From the point of view of the calculation, the integral sign is replaced by a summation sign. The $r$ and $\theta$ terms are then $r_i$ and $\theta_i$.

also, $E = AF - C^2$

and $F = \int_{\theta_1}^{\theta_2} \cos^2 \theta \, d\theta - \frac{1}{\theta_2 - \theta_1} \left( \int_{\theta_1}^{\theta_2} \cos \theta \, d\theta \right)^2$

(11.9)

## 11.9 Other parameters

These are:
(1) $dr/d\theta$.
(2) Curvature.
(3) Harmonic analysis.

### 11.9.1 dr/dθ

This is the rate of change of radius (See Figure 11.51). Usually, the average value of $dr/d\theta$ over a small window of angle is used, the angle being about 5°, or sometimes the maximum value of $dr/d\theta$ is taken around the circle. There are two points to notice. One is that this does not relate to the out-of-roundness error and the other is that it is not sensitive to the centre position of the roundness data. $dr/d\theta$ was introduced by ball and roller bearing makers who were trying to correlate roundness parameters with the acoustic noise generated by some bearings. There are two reasons put forward to justify $dr/d\theta$ rather than roundness parameters. The first is that the ear responds to velocity i.e. momentum change. The other more likely reason is that it is the $dr/d\theta$ that imparts impulses to the races as the bearing turns, thereby causing vibration. It is used extensively in miniature precision bearing manufacture. The roundness value is often referred to as DFTC (**departure from true circle**) in these firms.

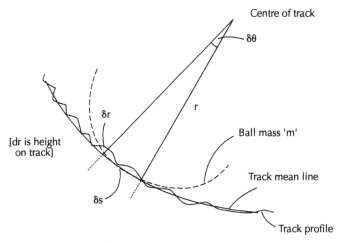

**Figure 11.51** *Significance of dr/dθ*

**Note**: $mr$ = impulse on track normal to motion

= and if $w$ is the angular velocity

$$m \frac{dr}{dt} = m \frac{dr}{d\theta} \cdot \frac{d\theta}{dt} = m \frac{dr}{d\theta} w \qquad (11.10)$$

### 11.9.2  Curvature

This is taken as measuring the radius of curvature of a nominally circular arc, or sometimes the change in curvature or the centre of curvature. For this parameter, the value of $d^2r/d\theta^2$ is needed: the rate of change of $dr/d\theta$. Curvature and to some extent $dr/d\theta$ (or slope) are susceptible to noise and therefore have to be evaluated with care. It is unfortunate that the noise may sometimes be the actual roundness data itself. The roundness data can, in such circumstances, produce a bias on the value of the curvature, which is itself determined from the same data. Errors of 5% are not uncommon. The $dr/d\theta$ and curvature parameter are differentials of the roundness data. Curvature can often be used in industry. One example is in bar-relled roller bearings (Figure 11.52).

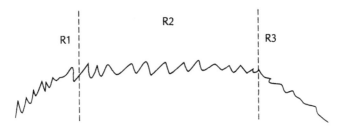

**Figure 11.52**   *Curvature required at three places*

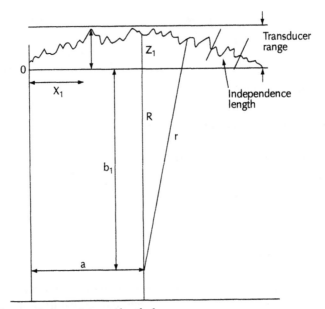

**Figure 11.53**   *Form and texture – integrated method*

The $R$ values are obtained as the reciprocal of the curvature

$$R = 1 \bigg/ \sqrt{\frac{d^2r}{d\theta^2}} = \frac{1}{C} \tag{11.11}$$

To get good results e.g. accuracy to $1:10^3$, it is necessary to use good numerical analysis procedures. In some circumstances, curvature can be derived from a surface texture graph. Figure 11.53 shows a graph taken on a surface texture instrument having a wide range. The radius of the part i.e. the reciprocal of the curvature can be evaluated at the same time as the texture. This method has the advantage that only one calibration is needed for both the curvature and the texture. Also, there is only one setting-up procedure required (problems with this are illustrated in Figure 11.54).

### 11.9.3  Harmonic analysis

It has already been shown in 11.5.1 that Fourier analysis is useful in analysing the various methods of roundness measurement. It is also useful for characterizing the roundness signal.

In this, the data representing the roundness signal from the workpiece is analysed in the conventional way to find the 'spectrum' of the data e.g. [11.6]. Roundness data is especially suited to harmonic analysis because it is repetitive and the Fourier components representing the Fourier coefficients of the series use this repetition to good effect. Notice that it is more difficult to analyse surface roughness by this means because there is no given periodicity. Obviously, the set of Fourier coefficients representing the roundness data constitute more than one parameter. In fact, the series of coefficients can be used very effectively. Thus, starting at the low upr and moving to the higher upr enables many factors of roundness measurement to be investigated.

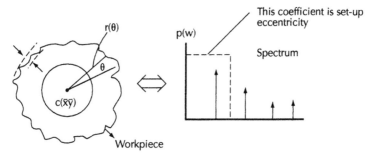

**Figure 11.54**   *Set-up problems*

$$P(n \text{ lobes per rev}) = \sum r(\theta) \exp \frac{(jn\theta)}{T} \qquad (11.12)$$

### Fourier coefficient

$n = 0$ corresponds to the radius of the workpiece that is suppressed by moving the probe to touch the workpiece.

$n = 1$ is eccentricity of the centre of the workpiece relative to the centre of rotation.

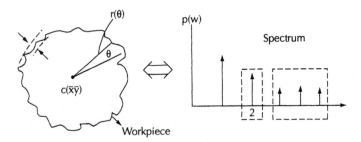

**Figure 11.55**　*Machine problems*

$$P(n \text{ lobes per rev}) = \sum r(\theta_i) \exp \frac{(jn\theta)}{T} \qquad (11.13)$$

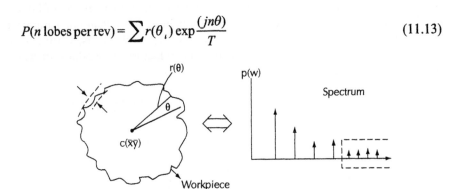

**Figure 11.56**　*Process or material problems*

$$P(n \text{ lobes per rev}) = \sum r(\theta) \exp \frac{(-jn\theta)}{T} \qquad (11.14)$$

### Process or material problems (Figures 11.55 and 11.56)

$n = 25–100$: process faults – built-up edge in turning, wheel loading in grinding.

$n = 100–1000$: usually material problems.

Most of the problems can be attributed to a specific band of spatial frequencies.

The coefficients of lobing are evaluated as below:

$$p(\theta) = R + \sum_{n=1}^{a} C_n \cos(n\theta - \varphi_n)$$

$$= R + \sum_{n=1}^{a} (a_n \cos n\theta + b_n \sin n\theta) \qquad (11.15)$$

$$C_n = \left(a_n^2 + b_n^2\right)^{\frac{1}{2}} \text{ and } \varphi_n = \tan^{-1}(b_n/a_n)$$

$$R = \frac{1}{2\pi}\int_{-\pi}^{\pi}(\theta)d\theta \text{ or in digital form } \frac{1}{N}\sum r_1(\theta)$$

$$a_n = \frac{1}{\pi}\int_{-\pi}^{\pi} r(\theta)\cos n\theta \ d\theta \text{ or } \frac{2}{N}\sum_{i=1}^{N} r_1(\theta)\cos n\theta \qquad (11.16)$$

$$b_n = \frac{1}{n}\int_{-n}^{\pi} r(\theta)\sin n\theta \ d\theta \text{ or } \frac{2}{N}\sum_{i=1}^{N} r_1(\theta)\sin n\theta$$

$$C = \frac{2A}{2}\sin\left(\frac{n\alpha}{2}\right)$$

## Summary

| Coefficient 0 | Represents size. |
| 1 | Represents eccentricity i.e. instrument set-up. |
| 2 | Ovality. |

| Coefficients 3-5 | Distortion of the work by clamping during manufacture. |
| 6-20 | Chatter caused by lack of rigidity of the machine tool. |
| 20-100 | Process effects, tool mark – feed built-up edge etc. |
| 100-1000 | Material effects i.e. microfracture. |

In other words, bands of coefficients can be used to monitor instrument set-up, workpiece set-up, machine tool, process and material effects. All of these are taken from one workpiece. This realization led to the idea that the workpiece could be represented as a 'fingerprint' of the whole of manufacture.

The example above is only for one roundness profile. The information from the whole of the workpieces to include cylindricity straightness etc. enables more information about the manufacture to be obtained.

## 11.10 Filtering for roundness

The filters used in roundness had their characteristics decided by use (Figure 11.57). It is no coincidence that the banding of the harmonics (lobes) follows closely the filter characteristics upr (undulations per revolution).

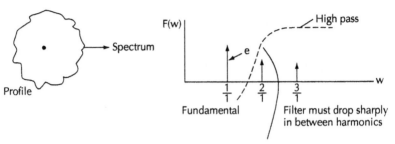

**Figure 11.57** *Filter requirement for roundness*

Thus, typical filters are:

1 to 15 upr;

1 to 50 upr;

1 to 500 upr;

15 to 150 upr;

15 to 500 upr.

Nowadays, there are, in addition, flexible cut-offs to allow the customer to choose filters for their own particular use. Also, as in roughness, 2CR filters (phase corrected) and Gaussian filters would be available. See, for example, ISO 4291 1985.

The filters used in roundness measurement show the characteristics that have evolved mainly because of problems in manufacture. Some of the ranges have also been found to be useful in functional situations, especially in the bearing industry. There is, however, growing interest in having filters that can match specific function requirements. These filters are **functional filters**. An example is shown in Figure 11.58.

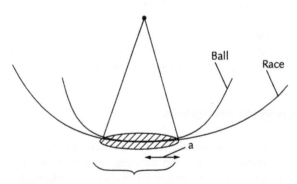

**Figure 11.58** *Functional filtering*

Knowing the ball loading $W$, the geometry of the bearing $R$ and $E$ the elastic modulus, the Hertzian elastic radius is found. This value can be used to determine the filter cut-off from the radius 'a' of the contact,

$$a = 1.11 \left( \frac{WR}{E} \right)^{\frac{1}{3}}$$ 

(11.17)

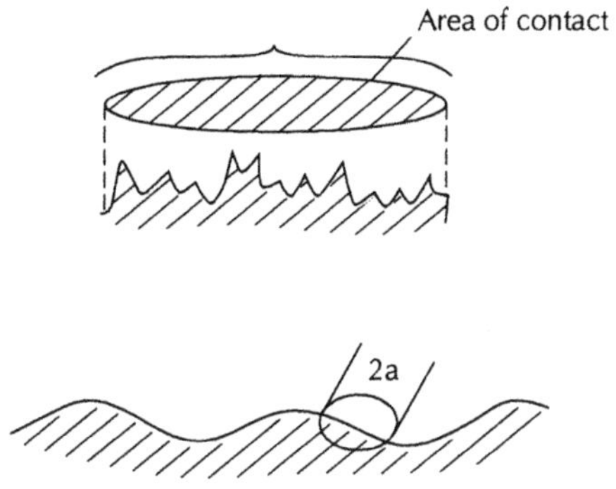

Figure 11.59   *Determining the filter cut-off*

The small asperities will be within the Hertzian zone and could be a source of acoustic noise. For functional acceptance, the high pass filter should be at $\frac{1}{2a}$ as shown in Figure 11.59.

The low pass filtering at $\frac{1}{2a}$ gives the waviness which tends to give acceleration to the ball as a whole and so be a source of vibration of the bearing.

## 11.11   Harmonic problems

Having extolled the value of harmonic analysis, it should be noted that it has its problems. One of these problems is that parts of the roundness signal are not amenable to harmonic analysis. This is when there is a spatial feature. Sometimes this is best left as a feature rather than breaking it down into harmonics.

For example, a filter to remove the eccentricity is expected to remove $A_1\cos\theta$, where $A_1$ is the magnitude of the eccentricity. What the filter does is to remove *any feature that happens once per revolution*. This is not the same as *one undulation per revolution*.

Consider Figure 11.60 (a), which is a nominally circular part having a dimple or flat over one part of the arc. The reference line through the figure after having removed the 1 upr signal is shifted. This result is not wrong – the departures from the reference line are real as seen in Figure 11.60 (b), but Figure 11.60 (c) shows the practical realization in which the dimple feature is retained and the reference line follows exactly the arc of the circle, which is not what the Fourier analysis does.

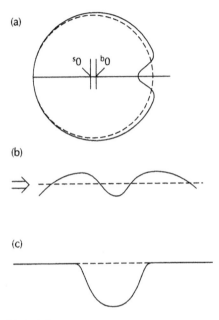

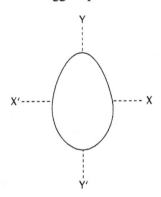

**Figure 11.60**   *Effect of atypical indentation*

Also consider what happens to an egg shape as shown in Figure 11.61:

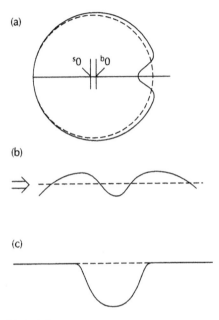

**Figure 11.61**   *An egg shape*

This has (i) an axis of symmetry YY'. Therefore, it has a coefficient of second harmonic – second lobe (ovality). It has (ii) an axis of asymmetry XX'. Therefore, it has a coefficient of first harmonic – eccentricity. This is not all clear from the Fourier analysis.

## 11.12   Alternatives to harmonic analysis

### 11.12.1   General

There have been alternatives to the straightforward Fourier analysis. One is the average wavelength and another is the root mean square, given below.

**First harmonic problems**

$$C_1 \frac{2A}{\pi} \sin(\alpha / 2)$$  (11.18)

This has been discussed above.

**Average coefficient**

$$N_a = \frac{\sum\limits_{n=m_1}^{n=m_2} n(a_n \sin n\theta + b_n \cos n\theta)}{\sum\limits_{n=m_1}^{n=m_2} (a_n \cos n\theta + b_n \sin n\theta)}$$  (11.19)

**Root mean square**

$$N_q = \left( \sum_{n=1}^{m} (a_n^2 + b_m^2) n^2 \middle/ \sum_{n=1}^{m} (a_n^2 + b_n^2) \right)^{\frac{1}{2}}$$  (11.20)

### 11.12.2   Average wavelength

The average wavelength is an average parameter similar to the $\lambda_a$ value in roughness. As can be seen from the formula (11.19), it does not represent just wavelength on the surface, it weights each wavelength by its amplitude and presents a weighted wavelength, which is meant to be very stable and a good representation of the waves (upr) on the surface. Some examples are shown in Figure 11.62.

The average wavelength can be used in addition to the peak height values shown in conventional roundness assessment.

Just compare the profiles of the specimens shown in Figure 11.63. Each one of these has exactly the same value of out-of-roundness yet they are clearly different shapes. Obviously, the spectral analyses of these would be different but that

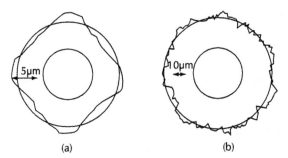

**Figure 11.62**   *Average wavelength for roundness (a) average height 0.3µm, average upr 4; (b) average height 1µm, average upr 20*

requires some considerable calculation. The average wavelength or average wave number takes account of the whole spectrum – the amplitude and wavelength of each of the components making up the profile graph – and so would be a straight-forward parameter to add to the roundness value [11.7].

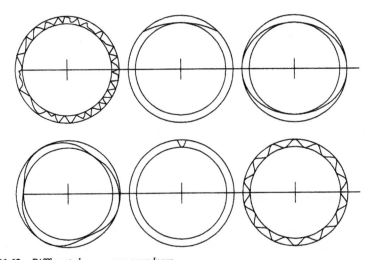

**Figure 11.63**   *Diffferent shapes, same roundness*

The idea is to take into account any very-high upr values that result from the edges, say, on a keyway, finding the average wavelength (or more accurately its recipro-cal $N_a$) and then high pass filtering after it. Note that the edges can have virtually a continuous spectrum and are hardly tied to the harmonics (Figure 11.64).

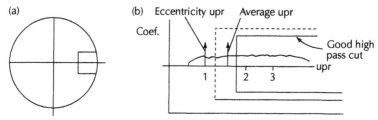

**Figure 11.64** *Broadband effect of keyway*

## 11.13 Non-roundness parameters

It has been pointed out earlier that the rotating table method for measuring roundness is the most popular radius measuring instrument because of being able to measure eccentricity etc. relatively easily. This makes it more suitable for cylindricity measurement. Below are shown some examples of non-roundness parameters. Figure 11.65 shows the theoretical and practical eccentricity values. Figure 11.66 shows concentricity as a function of eccentricity. Figure 11.67 shows ovality and Figures 11.68 and 11.69 show squareness and the flatness of specimen end. Figure 11.70 shows misalignment of axes. Figure 11.71 shows that it is possible to measure concentricity with a rotating stylus method. The one method shown uses a tree stylus. The concentricity between S₁ and S₂ are different. They depend on the 'l' values so some calculation is necessary to compensate this. If the spindle is moved vertically, it has to be done with great care because the spindle axis could move relative to the axis of the part, which will ruin eccentricity and hence concentricity values.

Figure 11.65 shows how concentricity can be measured with ease using the rotating table method. Moving the stylus from $S_1$ to $S_2$ does not change the position of the specimen axis with respect to the axis of rotation.

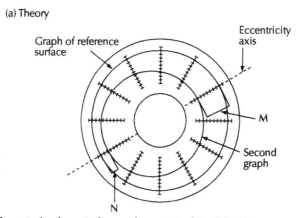

**Figure 11.65** *Theoretical and practical cases of eccentricity (M – N)/2.MAG*

(b) Practice

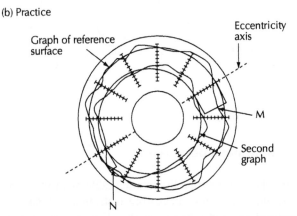

Graph of reference surface

Eccentricity axis

M

Second graph

N

**Figure 11.65 (continued)** *Theoretical and practical cases of eccentricity (M – N)/2.MAG*

## 11.13.1 Concentricity

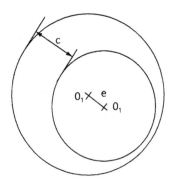

**Figure 11.66** *Concentricity, c*

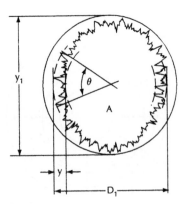

**Figure 11.67** *Practical case of ovality*

## 11.13.2   Squareness

It is a fact that roundness instruments are versatile enough to measure features that are not concerned with roundness. Perhaps the most useful is squareness. In this, the gauge is adjusted to measure in a plane perpendicular to the accurate spindle as shown in Figure 11.68. In this configuration, deviations recorded on the chart are deviations of the component in the axial direction rather than the radial direction. What appears to be eccentricity is in fact the y coordinate of the tilt of the part (seen where the problem is). Tilt is found by knowing the radial distance from the axis of rotation and dividing this into the 'y' (eccentricity) value. Squareness estimation relative to another feature having the same axis is now possible providing that it has been centred before measuring the tilt. Under these circumstances the tilt is the squareness of the one feature relative to the other.

A word of caution here! It has to be remembered that the accuracy of the spindle providing the reference rotation for roundness is not necessarily the same when it is being used for measuring squareness. There is no guarantee that the stiffness in the axial direction is as good as it has to be in the radial direction, which is necessary for roundness measurement.

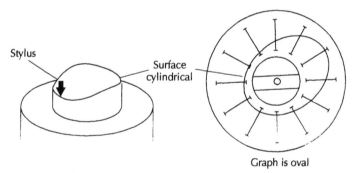

**Figure 11.68**   *Measurement around a horizontal surface*

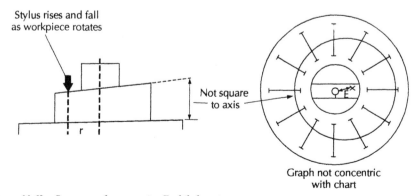

**Figure 11.69**   *Squareness by measuring E while knowing r*

### 11.13.3　Alignment of axes

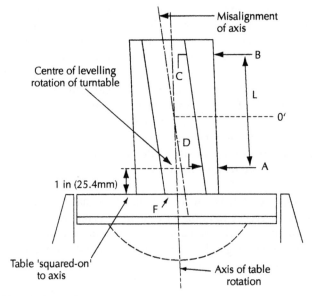

**Figure 11.70**　*Misalignment*

To determine the amount of misalignment, it is necessary to define a reference axis from the outside surface and align this axis to the axis of rotations of the turntable in such a way that the profile graphs from surfaces at A and B are concentric. Misalignment of the bore axis may then be measured by transferring the stylus to the bore and taking graphs at C and D, although movement of the pick-up position along the component does not in any way affect the alignment of the component relative to the selected axis.

There is another problem associated with the measurement of tilt. This concerns tilt measurement of an angled cylinder.

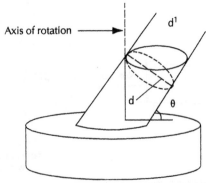

**Figure 11.71**　*Tilted cylinders*

Consider Figures 11.71 and 11.72. A trace of roundness taken parallel to the vertical axis as shown will look like an ellipse.

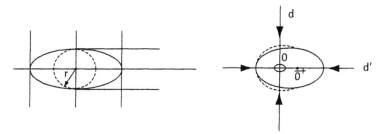

**Figure 11.72**   *Effective tilt*

The major axes $d'$ relative to the minor $d - 2r$ gives the angle of tilt of the cylinder,
$\dfrac{d}{d'} = \sin\theta$ from which $\theta$ can be estimated. This is a valid measurement if the cylinder is centred at 0 as it is in Figure 11.70. However, if the cross section is eccentric
such as in Figure 11.71, then the minor diameter $d$ has to be measured with care
because of the limacon effect: $d$ has to be measured through the centre of rotation 0
and not halfway between the diameter $d'$ at $0'$ (Figure 11.72).

### 11.13.4   Other tilt/eccentric problems

Shapes are sometimes produced that are instrumental and are not due to the shape
or position of the workpiece. One such case is sometimes found when measuring a
part with two probes diametrically opposed (Figure 11.73).

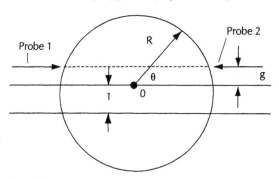

**Figure 11.73**   *Two probe problems*

Two probes are sometimes used to cancel out problems of eccentricity but they can
cause serious problems if the two-probe centre does not pass through the centre of
rotation at 0 but misses by an amount 'g'. If this system is used to measure a part
that is eccentric by 'e', then what emerges as the signal is:

$$\frac{l^2}{2R}(1-\cos 2\theta) + \frac{2eg}{R}(1-\cos \theta) \tag{11.20}$$

The ellipse (second term in equation (11.20)) is modified by the limacon shape produced at the same time and which add to it.

## 11.14  Conclusions

Roundness is one of the cornerstones of surface metrology for two reasons. One reason is that the roundness of a workpiece is directly linked to the performance of the machine tool: it is more relevant than roughness in this respect. The other reason is that roundness is the prerequisite of rotation, which in turn dominates dynamic function. It could be argued that roundness is to dynamic function what roughness is to contact. Rotation is the preferred form of energy transfer because it is localized to the centre of rotation whereas translation forms are not.

Roundness instruments form the basis for measuring many metrological features apart from roundness such as :
1. Concentricity.
2. Squareness.
3. Alignment.
4. Cylindricity.
5. Sphericity.
6. Conicity.

These make up many of the metrology requirements for the automotive industry, for example.

Partial roundness generates its own family of important parameters such as $dr/d\theta$ and curvature.

There are some problems associated with roundness. One of these is the persistent use of peaks and valleys as one of the bases of roundness characterization. These are in the form of plug, ring and minimum zone methods. It is a mystery why these peak orientated definitions have not fragmented into average peaks in the same way that $R_y$ was largely replaced by $R_z$ in roughness. It is to be hoped that the least-squares method will remain the dominant tool.

In miniaturization, there is a growing problem in measuring roundness. It is much more difficult than measuring the roughness of small components. It is likely that a complete instrumental rethink is required here.

# References

11.1    Reason R. E. *Report on the Measurement of Roundness*. Rank Precision Industries Leicester (1966).

11.2    Eisenhart L. P. A. *Treatise on the Differential Geometry of Curves and Surfaces*. Dover New York (1909).

11.3    Witzke F. *In Situ Out of roundness*. Proc. I. Mech. E, Vol. 182 pt. 3K p430 (1967).

11.4    Whitehouse D. J. *Radial Deviation Gauge*. Precision Engineering Vol. 4 p201 (1987).

11.5    Whitehouse D. J. *A Best Fit Reference Line for Use in Partial Arcs*. J. Phys E. Sci. Instr. Vol. 6 p921 (1973).

11.6    Papoulis A. *Fourier Integral and Application*. McGraw-Hill New York (1962).

11.7    Spragg R. C. and Whitehouse D. J. *A New Unified Approach to Surface Metrology*. Proc. Inst. Mech. Eng. Vol. 185 p697 (1970–1971).

# 12

## Cylindricity, sphericity

### 12.1 Cylindricity

The natural extension to roundness and its measurement is sphericity, but in fact easily the most important solid extension is cylindricity. The fact is that the need to generate two rotations is small. A cylinder, on the other hand, allows the possibilities of rotation and/or translation. This is the combination of the circle with a straight line generator.

For many years after roundness instruments had been developed by Reason, the measurement of cylindricity seemed unnecessary. The reason given now sounds naïve: 'If the shaft is circular in three places and the surface finish is acceptable the shaft is bound to be a good cylinder' (Figure 12.1).

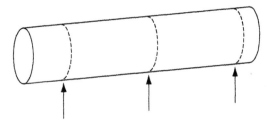

**Figure 12.1**  *Roundness checks*

The assumption was that only a really good machine tool would produce a circular and smooth shaft and by inference a good cylinder. Now cylindricity should be taken more seriously. The problem is that it is complicated.

One look at a crankshaft, for example, is enough to be convinced of the need to measure and to correlate profiles in different planes. A typical situation is shown in Figure 12.2.

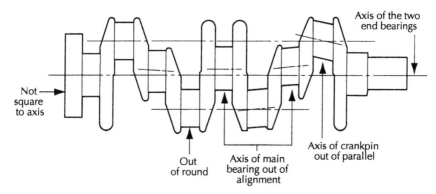

**Figure 12.2**   *Possible errors in crankshaft geometry*

There are some basic form errors and positional errors that occur regularly in cylindricity. Some are shown in Figure 12.3.

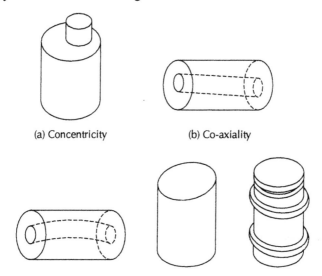

(a) Concentricity          (b) Co-axiality

**Figure 12.3**   *Errors of form that can be checked*

Eccentricity and concentricity are two of the most important parameters. The assessment is shown in theory and practice in Figure 12.4.

Eccentricity assessment = $(M − N)/2 × 1$/magnitude, where M and N are in inches or millimetres.

(a) Theory

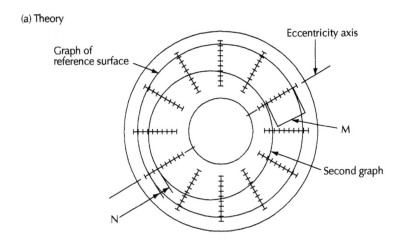

Graph of reference surface

Eccentricity axis

M

Second graph

N

(b) Practice

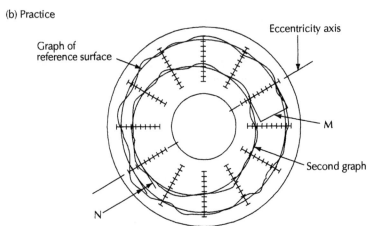

Graph of reference surface

Eccentricity axis

M

Second graph

N

**Figure 12.4**   *Eccentricity assessment (as in Figure 11.65)*

Figure 12.5 illustrates some of the shapes that would nominally come under the heading of cylinders. Those in (a) have bends in the axis. Those in (b) have radial variation along the axis, while those in (c) have distortions within the profiles of the cylinder.

Figure 12.6 shows that there are a confusion of shapes that need to be considered as cylinders. These shapes are highly exaggerated.

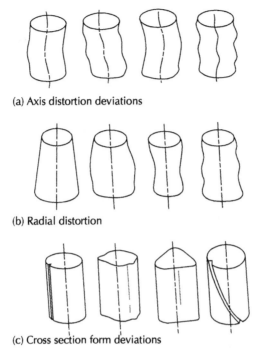

(a) Axis distortion deviations

(b) Radial distortion

(c) Cross section form deviations

**Figure 12.5** *Three types of error in cylindrical form*

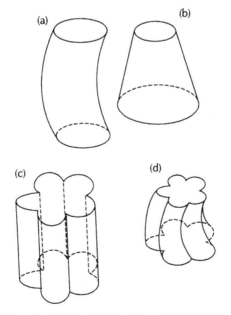

**Figure 12.6** *Errors in cylindrical form; basic types of deviations*

## 12.2   Configurations for measurement

Figure 12.7 shows some ways in which a cylindrical form could be measured. Figure 12.7 (a) is probably the obvious one. The measurement would be basically of roundness, linked with some straightness measurement to consolidate the cylindrical form.

This approach is sensible because it fits in with the way that roundness instruments, especially the rotating table type, operate. The critical values are the roundness and the associated centres at each level.

The rotating table instruments have a column upon which the stylus arm is mounted and which usually has good straightness characteristics. See Plates XIII and XIV.

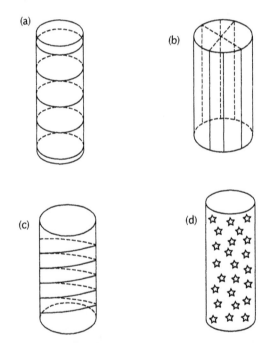

**Figure 12.7**   *Methods of measuring cylinders*

The other possibility (Fig. 12.7 (b)) is not preferred because it puts too much emphasis on the axial generator, which is usually less controlled than roundness. A popular possibility is the spiral mode. This has the considerable advantage of being able to take all the measurements without stopping the table or the column movements, so potentially it is a good method, but it mixes up the two coordinate systems, each of which can be broken down into a Fourier series. Unfortunately, the length of a cylinder does not comfortably break into a series but into a continuous Fourier transform. In effect, understanding the signal produced by a roundness instrument for a cylindrical part can be difficult. This is because the linearization of the circle formula to get a limacon for radial signals does not extend to the axial signal. The discrete spot method is the complete opposite of the spiral trace. At present, it is not an option because of the time taken to make a complete measurement, although this method has the advantage that it can be used by a CMM.

Figure 12.8 shows some of the methods of assessing the 'cylindricity'. The figures (a) to (d) follow on from the standard methods of measuring roundness, namely least squares (a), ring gauge (b), plug gauge (c) and minimum zone (d).

The scheme usually adopted is the least-squares method (Figure 12.8 (a)). In this, the cylinder axis is usually generated by working out the least-squares centres of all the polar graphs. The best-fit line is found from the centres and then the deviations at each level found from the cylinder generated about this axis. The cylindricity is usually given by the maximum peak deviation added to the biggest valley deviation from the least-squares radius of all the data.

Of the other methods (b), which is based on the ring gauge (minimum circumscribing circle), is probably best because it is unambiguous. It is just a minimum collar around all the data. Once this has been found using the points scheme in Tables 10.2 and 10.3, the cylindricity can be determined. It is the deepest valley from the collar. Cylinders derived from (c) plug gauge and (d) the minimum zone are, to some extent, not unique. An irritating point is that the drawing codes specify a minimum zone cylinder as the preferred method but do not show how to do it!

There are two other methods of assessing the cylindrical shape. The one by Goto [12.1] (Figure 12.9) has least squares at each level with its associated harmonic analysis. The best fit axis is a polynomial fit.

This is not altogether arbitrary. There is no technical justification for keeping the least-squares method for both radial and axial directions. The method has not been adopted because of its complexity. Incidentally, there is a definite reluctance of workers to use more than one or two numbers. 'One figure cylindricity' is an aim which has not yet been fulfilled!

One very comprehensible method relies on the least-squares approach but the result is broken down into its essential components. The radius term (Figure 12.10 (a)) and the coordinate of the centre for each profile (Figure 12.10 (b)) are plotted as a function so that two independent graphs can be drawn. These two graphs are a function of height z along the cylinder axis and highlight very easily problems in manufacture; (a) can indicate squareness, (b) can indicate slide error, and so on. In principle, the same method could be used on any of the gauge-like roundness assessments as well as least squares.

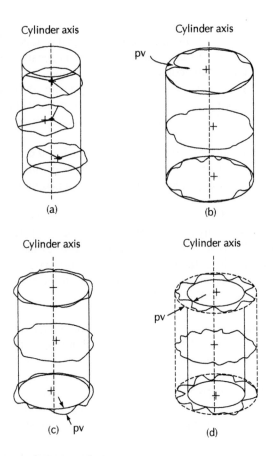

**Figure 12.8**   *Methods of defining a cylinder*

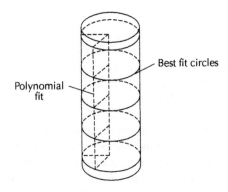

**Figure 12.9**   *Cylindricity method by Goto*

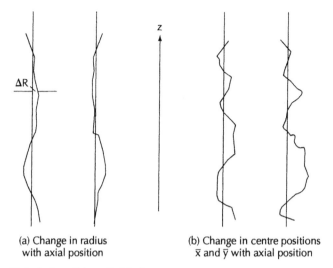

(a) Change in radius
with axial position

(b) Change in centre positions
$\bar{x}$ and $\bar{y}$ with axial position

**Figure 12.10**  *Method of specifying cylindrical error*

**Note 1**: often the best-fit least-squares cylinder is derived from all the collected data. This is called the **referred cylinder**. The errors in cylindricity are then taken as the deviations in peak and valley from it using the original data.

**Note 2**: quite often, the minimum zone cylinder is specified on drawings yet gives no indication of how to obtain it. Linear programming methods such as the simplex [12.2] technique can be used but there is no international agreement. The same is true for tolerancing. Fitting a cylinder representing a shaft into another cylinder representing a bore presents problems in tolerance specification, even with vector tolerancing methods now being adopted [12.2].

There is one other method for describing the cylinder and this is equivalent to specifying Cartesian coordinates for roundness. This is the method where the surface of the cylinders is developed i.e. put on a plane. Figure 12.11 (a) shows the development of a cylinder and Figure 12.11 (b) that for a cone. For this display, there are usually as many generator (straightness) measurements as there are roundness graphs. The coverage is therefore like a net that can be clearly seen when the cylinder has been developed. Having this 'net' arrangement makes it possible to optimise the geometry because the least-squares deviations can be evaluated at each node. This means that a noise can be minimized by going around each loop completely and checking for consistency.

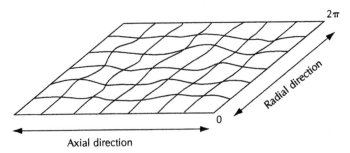

**Figure 12.11 (a)**   *Development of cylinder surface*

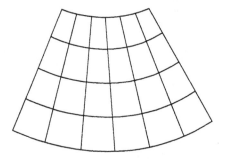

**Figure 12.11 (b)**   *Conality development*

The development method is quite useful for measuring and viewing wear scars but has not been widely accepted because it has lost the visual impact of the cylindrical coordinate system retained by the other methods.

### Tilted cylinders

Should a cylinder be tilted when measured, the result will appear as an ellipse. Therefore, it is essential that any levelling of the cylinder is performed before measurement. However, it may be that if a second cylinder is being measured relative to the first (e.g. for coaxiality), re-levelling is not practical (since the priority datum – the first axis will be lost). In this case, it is necessary to correct for tilt by calculating the tilt and orientation of the axis and noting the radius of the second cylinder, and to compensate by removing the cylinder tilt-ovality term for each radial plane prior to performing the cylinder tilt.

Removal of the second harmonic term or applying a 2 upr filter is not the correct procedure because any true ovality in the component will also be removed.

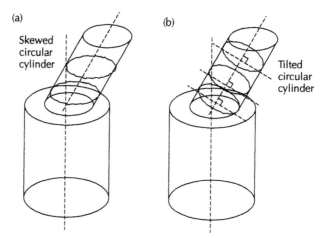

**Figure 12.12**   *Effect of tilting cylinder*

## 12.3   Some definitions of cylindrical parameters

### 12.3.1   *Coaxiality*

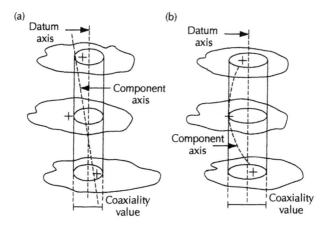

**Figure 12.13**   *Coaxiality*

If a best straight line axis is used using the chart eccentricity centres, the maximum deviation of this derived line from the datum line determines the radius of the cylinder around the datum axis. This is called the coaxiality of the 'three' charts (Figure 12.13 (a)). If a curve is filtered through the eccentric centres, then the radius of coaxiality is the maximum deviation of this deviation axis from the datum axis (Figure 12.13 (b)).

### 12.3.2 Asperity effect

All the zonal methods of assessing cylindricity are prone to errors if there is an abnormal peak on any of the roundness profiles. It is often acceptable to ignore a rogue peak (Figure 12.14).

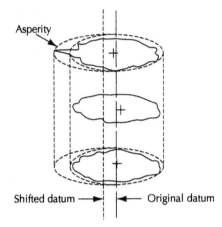

Asperity

Shifted datum ⟶  ⟵ Original datum

**Figure 12.14** *Asperity effect*

### 12.3.3 Ambiguous fits

Fitting cylinders to other shapes e.g. a cone can cause a serious ambiguity (Figure 12.15). This is similar to the plug gauge problem in roundness.

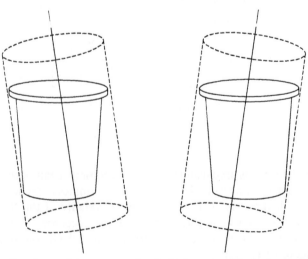

**Figure 12.15** *Some factors that cause errors in cylindricity – mismatch of reference shape to test object*

### 12.3.4    Run-out

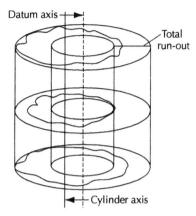

**Figure 12.16**    *Total run-out*

Total run-out is similar to 'total indicated reading' (TIR) or 'full indicated move-ment' (FIM) as applied to a roundness or two-dimensional figure, but in this case it is applied to the complete cylinder and is given as a radial departure of two concen-tric cylinders, centred on a datum axis that totally encloses the cylinder under test (Figure 12.16).

## 12.4    Assessment

The maximum peak to valley that decides the minimum concentric cylindrical zone is often used as the measure of the cylinder. Equally, harmonic analysis can be used. It depends on the application. Usually, the zonal fit is for static function and the least-squares cylinder for dynamic function.

## 12.5    Other points on cylinders

Figure 12.12 shows how difficult it is to measure objects with multiple cylinders. The problem is that if the two cylinders are at angles to each other as seen in the picture, there will be interactions between them and the tendency to produce ellip-ses on the tilted cylinder, with limacons produced by eccentricity or tilt on the major cylinder. Elliptical and limacon distortions are, unfortunately, at right angles and can almost cancel out.

What features of the cylinder shape are most important? Obviously, this depends on the application but there is a general rule of metrology when applied to function. This is that errors leading to asymmetry with respect to movement are the most important. Basically, where rotation is concerned, the axial problems are more important than the radial. Figure 12.17 (a) shows some possible axial shapes. The point is that any deviation of the axis from straightness is likely to cause vibration when rotating (Figure 12.17 (a)). Simply making sure that the cylinder passes the dimensional tolerance test is not enough. Radial deviations leading to size change are possibly more important in translation, causing asymmetry of properties e.g. the cylinder can act like a pump (Figure 12.17). Dimensional tolerancing can help satisfy the static requirements such as assembly but not the dynamic requirement. Zonal methods are dominated by the asperities on the surfaces and, because they are extreme, tend to be critical in assembly – the basic tolerance problem. In dynamic situations, the odd peak tends to rub away quickly and play no part in the function. It is the body as a whole that is important and the body or core of the surface geometry. This is where the least-squares criterion is best. It is the average or rms deviations that enter into the calculations [12.4].

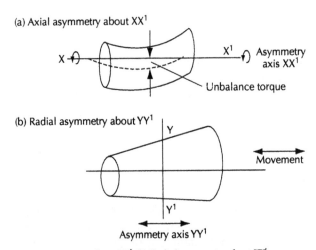

Figure 12.17   (a) Axial asymmetry about XX' (b) Radial asymmetry about YY'

Of the two effects, axial asymmetry is probably most important because rotation is usually the dominant motion.

## 12.6   Sphericity

Perhaps the most controversial aspect of sphericity is how it is measured. With roundness measurement, it is obvious that there is only one way to measure but with sphericity there are options. Great circles through the poles linked by the equator is one way. Latitude is another possibility, linked by great circles through the poles. The former is probably the best but it does not follow easily from a roundness set-up. The latter does but gets into the problem of probe angle at the poles. These are shown in Figures 12.18 and 12.19.

Despite being the natural extension of the circle, sphericity does not have many applications. Ball bearing measurement is an obvious one. Gyroscopes are another. More often encountered are partial spheres. These are found in the automotive industry in constant velocity joints. In medicine they are found, for example, in hip joint replacements.

In fact, having two or even three rotational axes as in goniometry is rare. It is possible to get two rotations in the same plane as in epitrochoids and some gears but again rarely. It is much more likely to get mixtures of translations and rotations as in cylindricity and conicity. The coordinate system usually used is spherical. This has two angles and one radius variable (Figure 12.18).

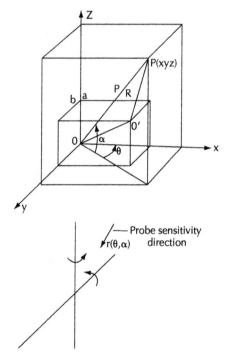

**Figure 12.18**   *Coordinate system for sphericity*

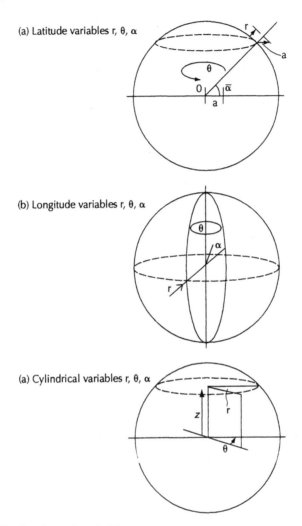

**Figure 12.19**   *Coordinates for sphericity*

## 12.7   Partial sphericity

As in most engineering problems, the real trouble begins when measuring the most complex shapes. Spheres are never, or hardly ever, complete. For this reason, estimates of sphericity on partial spheres – even full spheres – are made using three orthogonal planes. This alternative is only valid where the method of manufacture precludes local spikes. Similarly, estimates of surface deviations from an ideal spherical shape broken down in terms of deviations from ideal circles are only valid if the centres are spatially coincident – the relation between the three planes must be established somewhere! With components having very nearly a spherical shape, it is usually safe to assume this if the radii of the individual circles are the same [12.5].

In the case of a hip prosthesis (Figure 12.20), the difficult shape of the figure involves a further reorganization of data, because it is impossible to measure complete circles in two planes. In this case, the partial arc limacon method proves to be the most suitable. Similar problems can be tackled in this way for measuring spheres with flats, holes etc. machined onto or into them.

The display of these results in such a way as to be meaningful to an inspector is difficult, but at least with the simplified technique using orthogonal planes the three or more traces can all be put onto one polar or rectilinear chart. Visually, the polar chart method is perhaps the best, but if, for example, wear is being measured on prosthetic heads it is better to work directly from Cartesian records in order to measure wear (via the area under the charts) without ambiguity.

Assessment of the out-of-sphericity value from the least-squares centre is simply a matter of evaluating the maximum and minimum values of the radial deviations of the measured data points from the calculated centre $(a,b,c)$ and radius $R$. See Equation (12.1), (12.2) for partial zones.

The minimum zone method of measuring sphericity is best tackled using exchanged algorithms. Murthy and Abdin [12.6] have used an alternative approach, again iterative, using a Monte Carlo method which, although workable, is not definitive.

The measurement of sphericity highlights some of the problems that are often encountered in surface metrology, that is the difficulty of measuring a workpiece using an instrument which, even if it is not actually unsuitable, is not matched to the component shape.

If there is a substantial difference between the coordinate systems of the instrument and that of the component, artefacts can result, which can mislead and even distort the signal. Sometimes, the workpiece cannot be measured at all unless the instrument is modified. An example is that of measuring a spherical object with a cylindrical coordinate instrument. If the coordinate systems are completely matched, then only one direction (that carrying the probe) needs to be very accurate and sensitive. All the other axes need to have adjustments sufficient only to get the workpiece within the working range of the probe. This is one reason why the CMM has many basic problems: it does not match many shapes because of its versatility, and hence all axes have to be reasonably accurate.

The problem of the mismatching of the instrument with the workpiece is often true of cylindricity measurement, as will be seen. Special care has to be taken with cylinder measurement because most engineering components have a hole somewhere, which is often a critical part.

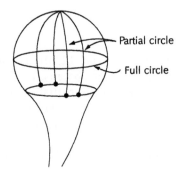

**Figure 12.20**   *Prosthetic hip head – partial sphere*

Full sphere:

In the case where $\theta_1 - \theta_2 = 2\pi$ and $\alpha_2$ and $\alpha_1 = 2\pi$:

$$a = \frac{4}{4\pi^2} \int_0^{2\pi}\int_0^{2\pi} r(\theta,\alpha)\cos\alpha \, d\theta d\alpha$$

$$b = \frac{4}{4\pi^2} \int_0^{2\pi}\int_0^{2\pi} r(\theta,\alpha)\sin\theta\cos\alpha \, d\theta d\alpha \qquad (12.1)$$

$$C = \frac{2}{4\pi^2} \int_0^{2\pi}\int_0^{2\pi} r(\theta,\alpha)(\theta,\alpha)\sin\alpha \, d\theta d\alpha$$

$$R = \frac{1}{4\pi^2} \int_0^{2\pi}\int_0^{2\pi} r(\theta,\alpha) \, d\theta d\alpha$$

Partial sphere

$$\theta_2 - \theta_1 = 2\pi, \alpha_2 - \alpha_1 = \delta\alpha$$

$$a = \frac{1}{\pi\cos\alpha} \int_0^{2\pi} r(\theta,\alpha)\cos\theta \, d\theta$$

$$b = \frac{1}{\pi\cos\alpha} \int_0^{2\pi} r(\theta,\alpha)\sin\theta \, d\theta \qquad (12.2)$$

$$R = \frac{1}{2\pi} \int_0^{2\pi} r(\theta,\alpha) \, d\theta$$

## 12.8   Wankel – other shapes

There are occasions where two rotations are needed as in the latitude and longitude methods in sphericity but the two rotations are in the *same plane*. The epitrochoid needed for the stator of the Wankel engine is a good example. Figure 12.21 shows how the locus of the minor circle rolls over the major circle and produces the required shape. For the hypotrochoid, the minor circle rolls on the inside of the major circle (Figure 12.21 (b)).

(a)

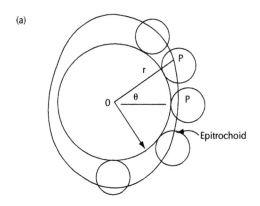

Measurement of Wankel motor stator

(b)

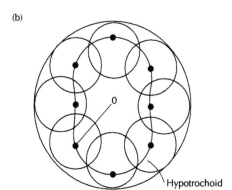

**Figure 12.21**   *Non-round polar forms*

## References

12.1 Iizuka K. and Goto M. Proc. Int. Conf. Prod. Eng. Res. Tokyo pt. 1 p431 (1974).

12.2 Chetwynd D. G. Ph. D. Thesis Leicester (1980).

12.3 Trumpold H. Tolerierung von Massen und Massketten im Austauschbau. VEB veriagtechnik, Berlin (1984).

12.4 Tonder K. *A Numerical Assessment of the Effect of Striated Roughness on Gas Lubrication.* Trans. ASME J. Tribol Vol. 106 p315 (1984).

12.5 Chetwynd D. G. and Siddall G. J. J. Phys. E. Vol. 9 p537 (1976).

12.6 Murthy T. S. R. and Abdin. *Minimum Zone Analyses of Surfaces.* J. M. T. D. R. Vol. 20 p125 (1980).

# 13

---

# Metrology instrument design and operation for minimum error

## 13.1 Introduction

Traditional surface finish is small when compared with dimension and the demands made on instruments have been extreme. This has required sophisticated mechanical design. This sophistication is not just for the instruments but also for investigating how surfaces function when in contact and moving.

The problem when investigating very smooth surfaces and measuring round and straight objects is that the instrument has to contain the reference from which the test surface is judged. So, the need to understand the surfaces has spawned the very shapes that were being examined, only much more accurately; hence the investigation of surfaces involved development of geometrics capable of providing flats, spindles etc. for the instruments. This involved the use of kinematics, and new materials.

Surface metrology instruments differ from machine tools because forces are small but accuracies have to be high. They vary from hand-held instruments to very large systems. They have to operate in environments that are relatively hostile: manufacturing is not the cleanest of operations. All eventualities have to be considered. Careful operation cannot be assured, so checks on performance and calibration are extensive as will be seen in this chapter and the next.

## 13.2   Capability

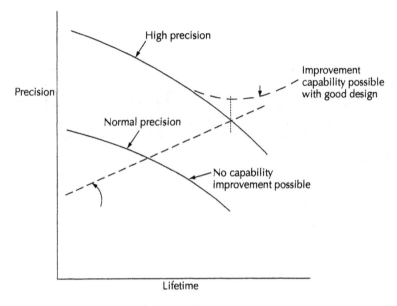

**Figure 13.1**   *Decline in instrument performance with time*

It is always the case that more is demanded of equipment/people/procedures than can reasonably be asked. One question often asked is whether an instrument's performance can be improved retrospectively i.e. after purchase and after use.

The quick answer usually is no, because the instrument has been designed to meet the specification and no more. It could be argued that over-design is bad design. However, requests for improvement will continue to be made.

One general rule of metrology is that it is not usually possible to make a bad instrument into a good instrument. It is possible, however, to make a good instrument better.

Good design improves the capacity to improve performance. Figure 13.1 shows the obvious, that all instruments will drop in performance with time.

The limiting performance is determined by the precision or repeatability of the system. The stability of the instrument determines the repeatability of measurement. The stability also determines how often the instrument needs to be calibrated (Figure 13.2).

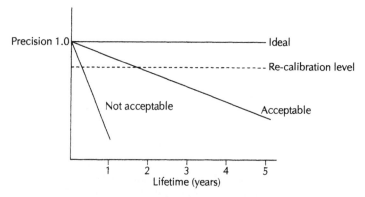

**Figure 13.2**   *Criteria for improvement in precision*

## 13.3   Errors [13.1]

Systematic error can be calibrated out of an instrument system as will be seen later in this chapter. The systematic error has to be that of a good instrument – the instrument has to be good enough to have a systematic error!

Random error is, by definition, not predictable and can only be got rid of by noise reduction paths or by averaging in time. Averaging can sometimes be worse than the noise problem because it allows time for other environmental errors to creep in.

## 13.4   Metrology instrument design and error reduction [13.2]

### 13.4.1   Introduction

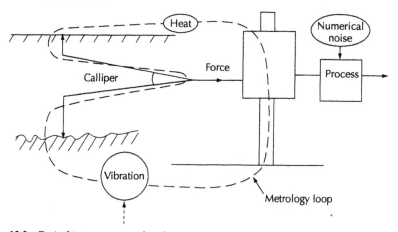

**Figure 13.3**   *Typical instrument metrology loop*

There are a number of major problems to be overcome.

1. Design problems.
2. Improving accuracy.
3. Set-up problems and usage.
4. Calibration and traceability.

The basis of all good instrument design is kinematics (Figure 13.3). These rules formulated by Whitworth and Kelvin not only give accurate results, they are usually cheap to implement. Some of the major points of design are given here. For a full exposition, see Smith and Chetwynd, and Moore [13.2, 13.3].

The basic questions that need to be addressed in any instrument or machine tool design are as follows.

1.   Is it correct according to kinematic theory, that is, does it have the required number of constraints to ensure that the degrees of freedom conform to the movements required?

2.   Where are the metrology and force loops? Do they interact? Are they shared and, if so, how?

3.   Where are the heat and vibration sources and, if present, how can they influence performance?

4.   Is the design symmetrical and are the alignment principles obeyed?

5.   What other sources of error are present and can they be reduced by means of compensation or nulling?

These points have been put forward to enable checks to be made on a design.

Some basic principles can be noted [13.4].

1.   Any unconstrained body has six degrees of freedom; three translational $x$, $y$, $z$ and three rotational $\alpha, \beta, \gamma$.

2.   The number of contact points between any two perfectly rigid bodies is equal to the number of constraints.

There a number of additional useful rules.

3.   Any rigid link between two bodies will remove a rotational degree of freedom.

4.   The number of linear constraints cannot be less than the maximum number of contacts on an individual sphere. Using the rules for linking balls and counting contacts, it is possible to devise a mechanism that has the requisite degrees of freedom. It should be noted that the definition of a movement is that which is allowed in both the positive and negative direction. Similarly, a constraint is such that

movement in one direction or both is inhibited. For example, it is well known that a kinematic design can fall to pieces if held upside down.

5.   A specified relative freedom between two subsystems cannot be maintained if there is under-constraint.

6.   Over-constraint does not necessarily preclude a motion, but invariably causes interaction and internal stresses if used and should therefore be avoided in instruments.

### 13.4.2   Kinematics

Kinematics in instrument design is an essential means to achieve accurate movements and results without recourse to high-cost manufacture (Figure 13.4). In essence, it is just a recognition of the fact that any rigid body in space has six degrees of freedom. Three of these are translational $x$, $y$ and $z$ and three are rotational $\alpha$, $\beta$, and $\gamma$. To stop a particular degree of freedom, there has to be constraint. So, for a body that is completely fixed, it has to have six constraints. One such example is the 'Kelvin clamp', described later. Below are examples of bodies having one constraint, two constraints … up to five. Then, some examples will be given.

One point needs to be made concerning constraints. Although a constraint in, say, the $x$ direction implies that no movement is possible either in the $+x$ or $-x$ direction, this is not necessarily so. In the design of kinematics, if constrained in $x$, a body can be restrained in $+x$ or $-x$ or both. In the first two of these cases $+x$ or $-x$ or the other direction $-x$ and $+x$ respectively are lightly held by springs or simply held in place by gravity. It is well known that kinematics can fall apart if held upside down. In practice, in machine tools and surface measuring instruments this is not usually a problem, although it could apply to hand-held devices.

| Example | Constraints | Degrees of Freedom |
|---|---|---|

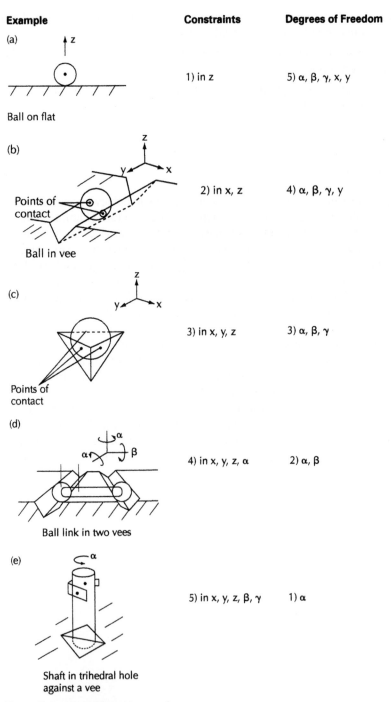

(a)

Ball on flat

1) in z

5) α, β, γ, x, y

(b)

Points of contact

Ball in vee

2) in x, z

4) α, β, γ, y

(c)

Points of contact

3) in x, y, z

3) α, β, γ

(d)

Ball link in two vees

4) in x, y, z, α

2) α, β

(e)

Shaft in trihedral hole against a vee

5) in x, y, z, β, γ

1) α

**Figure 13.4**   *(a)–(e) Kinematic examples*

Another simple system has a tube that needs to move linearly while resting on four balls (Figure 13.5). The tube has a slot in it into which a pin (connected to the frame of the instrument) is inserted to prevent rotation.

Pin to stop
rotation

**Figure 13.5** *Linear movement tube*

For one degree of rotational freedom, the drive shaft has a ball or hemisphere at one end. This connects against three flats, suitably arranged to give the three points of contact. The other two are provided by a vee made from two rods connected together (Figures 13.4 (c) and 13.7 (a)). Another version (b) is a shaft in two vees and resting on a flat. As mentioned earlier, these would have to be encouraged to make the contacts either by gravity or by a non-intrusive spring. There are numerous possibilities for such designs, which really began to see application in the 1920s.

In surface metrology instruments, kinematics is most often used in generating straight line movements or accurate rotations [13.3].

In both cases there have to be five constraints, the former leaving a translation degree of freedom and the latter a rotation. By far the largest class of mechanical couplings are for one degree of freedom. Couplings with more than one degree of freedom are seldom required, the obvious exception being the screw, which requires one translation and one rotation. Despite its apparent simplicity two translations cannot be achieved kinematically; $x$, $y$ slides are merely two systems of one degree of freedom connected together.

A typical translation system of one degree of freedom is shown in Figure 13.6. This has five pads on the carriage resting on the slide. The pads are usually polymeric. Figure 13.7 shows a single degree of rotation.

Side view 1          Side view 2          End view

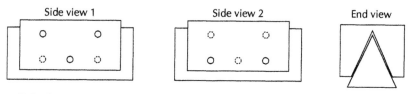

**Figure 13.6** *Kinematic translation system*

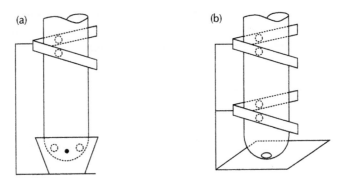

**Figure 13.7** *System showing single rotational degree of freedom*

The other main use of kinematics in surface metrology is in relocation. In this use, it is necessary to position a specimen exactly in the same position after it has been removed. Such a requirement means that the holder of the specimen (or the specimen itself) should have no degrees of freedom – it is fully constrained. The most usual arrangement to achieve this is the Kelvin clamp (Figure 13.8).

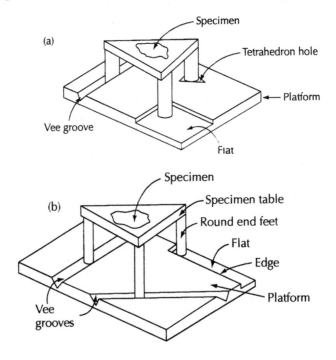

**Figure 13.8** *The Kelvin clamp*

The six degrees of freedom are constrained by three points of contact in hole C, which ideally is a tetrahedron. There are two points of contact in a vee groove at A and one on a flat at B. Another version is that of two non-parallel vees, a flat and an edge. Another version just has three vees.

The relocation requirement is usually needed when experiments are being carried out to see how the surface changes as a result of wear or some similar phenomenon. Also, comparison of performance of different instruments should use relocation to ensure that they all measure the same spot. See Figures 13.14–13.19.

Couplings with two degrees of freedom require four constraints. A bearing and its journal with end motion, Hooke's joint, the so-called 'fixed' knife edges, and four-film suspensions are common examples of this class.

One coupling with three degrees of freedom is the ball and socket joint. The tripod rests on a surface or a plane surface resting on another plane and three filar suspensions. Alternatively, the tripod could rest on a sphere in which only rotations are possible.

Above all, the hallmark of a kinematic design is its simplicity and ease of adjustment and, not insignificantly, the perfection of motion that produces a minimum of wear. In general, kinematic instruments are easy to take apart and reassemble. Kinematic designs, because of the point constraints, are only suitable for light loads – as in most instruments. For heavy loads, pseudo-kinematics is used in preference because larger contacts areas are allowed between the bearing surfaces and it is assumed that the local elastic or plastic deflections will cause the load to be spread over them. The centroids of such contacts are still placed in positions following kinematic designs. This is generally called pseudo- or semi-kinematic design.

### 13.4.3  Pseudo-kinematic design

This is sometimes known as elastic or plastic design.

Practical contacts have a real area of contact, not the infinitesimal area present in the ideal case in kinematics. As an example, a four-legged stool often has all four legs in contact yet, according to kinematics, three would be sufficient. This is because the structure flexes enough to allow the other fourth leg to contact. This is a case of over-constraint for a good reason, that is safety in the event of an eccentric load. It is an example of elastic design or elastic averaging.

Because in practice elastic or eventually plastic deformation can take place in order to support a load, semi-kinematic or pseudo-kinematic design allows the centroid and area of contact to be considered as the position of an ideal point contact. From this, the rules of kinematics can be applied.

In general, large forces are not a desirable feature of ultra-precision designs. Loads have to be supported and clamps maintained. Using kinematic design, the generation of secondary forces is minimized. These can be produced by the locking of a rigid structure and when present they can produce creep and other undesirable effects. For kinematics, the classic book by Pollard [13.4] should be consulted.

In general, another rule can be recognized, which is that divergence from a pure kinematic design usually results in increased manufacturing cost.

### 13.4.4   Hexapod considerations

Instruments and machine tools have to make the best use of the six degrees of freedom available to them. Robotic movement is complicated and the analysis complicated. Serial movement, although versatile, is liable to suffer from error propagation throughout the links. Hexapod systems based on the Stewart platform flight simulator [13.5] are now being developed. These are up to six movements in parallel and offer the possibility of instruments measuring complicated surfaces. Work is already in progress [13.6] on evaluating probe position and probe orientation for the workspace. Access to the working space is more limited than for serial movements but the potential for added stiffness is great. This is the trend in order to achieve sub-nanometre technology.

### 13.4.5   Mobility

The number of degrees of freedom is straightforward to work out with kinematics, but with the advent of parallel and serial machines and instruments it becomes more complicated. For this reason, the use of Grubler's method for determining movement is on the increase. This is in itself quite difficult but an example in two dimensions may help.

The mobility of a mechanism is given by:

$$M = 3(n-1) - 2f_1 - f_2$$

where $M$ is the mobility or degrees of freedom (DOF);
$n$ is the total number of linkages including the case or earth;
$f_1$ is the number of single degree of freedom joints;
$f_2$ is the number of two degree of freedom joints.

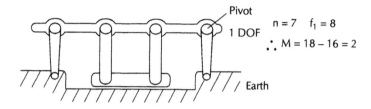

**Figure 13.9 (a)**   *Mobility of 2 degrees of freedom*

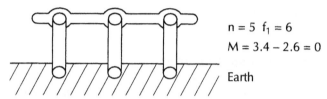

**Figure 13.9 (b)**   *Mobility of no degrees of freedom*

This answer seems (see Figures 13.9 (a) and (b)) surprising but it is true. All of the links to earth are of different lengths. There is no movement possible. These sort of calculations are invaluable if a number of linkages are free to move as in the hexapod systems. The three-dimensional version of this is especially useful.

### 13.4.6   Simple Abbé and cosine instrument error

The simplest case of Abbé error is where the scale of measurement is out of line with the object being measured. (Figure 13.10).

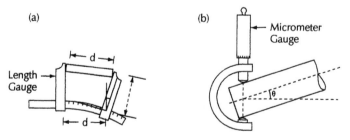

**Figure 13.10**   *Measurement errors*

Figure 13.10 (a) shows **Abbé error** in which the scale of measurement of the vernier is displaced by 'l' from the object. The d is obviously smaller than it should be

Figure 13.10 (b) shows the **cosine error** which occurs because the object is tilted with respect to the scale of the micrometer.

Abbé error is usually far more serious than cosine error. It is interesting to note that conventional CMMs (**coordinate measuring machines**) all violate Abbé's theorem. They minimize this by making the framework very stiff. The reason for the violation is to allow room for as many as possible different sized and shaped objects to be measured. In this case, accuracy is sacrificed for convenience and versatility.

Some instrument properties follow.

### 13.4.7  Metrology loop properties

The measurement loop must be kept independent of the specimen size if possible. In Figure 13.11, it is not. However, Figure 13.12 shows how it can be done.

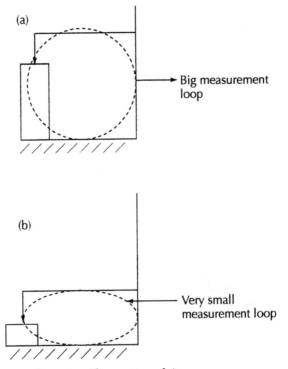

**Figure 13.11**   *Measurement loop – size with respect to workpiece*

In Figures 13.11 (a) and (b) the measurement loop is determined in size by the length of the specimen. Clearly, it is beneficial from the point of view of stability to have a small loop. This can sometimes curtail the ease of measurement, but this problem can be resolved in a completely different way as shown in Figure 13.12 below.

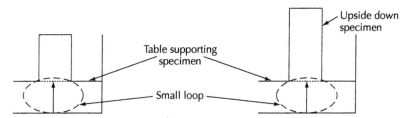

**Figure 13.12**   *Inverted instrument*

The measurement loop is independent of the size of the specimen. The measurement loop and also the force loop that moves the carriage etc. should be kept small. This is easy to see.

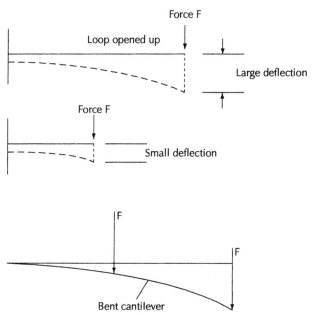

**Figure 13.13**   *Cantilever effect*

The small loop is effectively stiffer. The same force F produces a much smaller deflection e.g. consider a cantilever with a force on the end and the same force in the middle (Figure 13.13). The deflection itself is proportional to $l^3$, where $l$ is the loop length.

Small loops usually are beneficial (Figure 13.14) but sometimes make the operation of the instrument more difficult. Thin elements in the loop should be short.

Cross-sectional shape should be hollow, if possible like a tube.

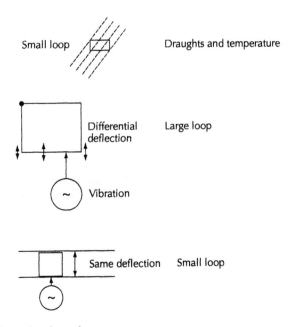

Figure 13.14   *Small metrology loop advantages*

The larger the loop, the more likely it is to be influenced by effects within the loop e.g. differentially by heat; all the small loop heats up so the shape does not change, but only part of the large loop will be heated (Figure 13.15). It is a similar case with vibration.

Figure 13.15   *Effects of loop size*

**Important note**: surface metrology has one big advantage over dimensional metrology. This is that it does not matter if the size of the metrology loop increases as long as the shape is preserved i.e. as long as the two arms of the calliper are in the same proportion. Consider the skid stylus system (Figure 13.16). The angle change $\delta\theta$ is the same for both systems.

Again, the smaller the loop, the less likely heat or noise sources will be within it.

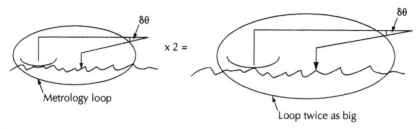

**Figure 13.16**    *Strict stylus system*

If there is a source within the loop, it is not serious providing that it is symmetrically placed relative to the two arms of the 'calliper'. The point is that any asymmetry has to be avoided.

### 13.4.8   Material selection and other problems

*Noise position.* If a noise source such as gearbox vibration is located, it should be removed or absorbed as close to the source as possible otherwise the noise can dissipate into the system via many routes as shown in Figure 13.17.

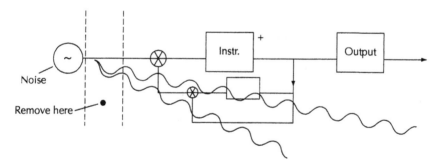

**Figure 13.17**    *Noise dissipation in system*

Material Selection [13.7]. Surface metrology, unfortunately, has to be able to measure very small detail and consequently must have short time stability. In fact, there are three physical properties that have to be considered where the material properties are concerned. These are:

1.    The ratio $E/\rho$ should be high so that the resonant frequency is out of the signal range. Also, I, the moment of inertia should be small ($E$ is Young's Modulus and $\rho$ the material density).

2.    Thermal response should be as fast as possible to dissipate thermal shock. The diffusion coefficient $\dfrac{K}{\rho\sigma}$ should be high where $K$ is thermal conductivity, $\sigma$ is the specific heat.

3.    Thermal conductivity $\dfrac{\alpha}{k}$ should be small for long term stability i.e. keep long-term expansion to a minimum. With low expansion materials such as Zeradur (fused silica and quartz), this is now possible.

Chetwynd [13.7] has examined a number of materials and plotted three very interesting graphs (Figures 13.18 (a), (b), (c)). The importance of silicon and silicon carbide as mechanical materials becomes obvious.

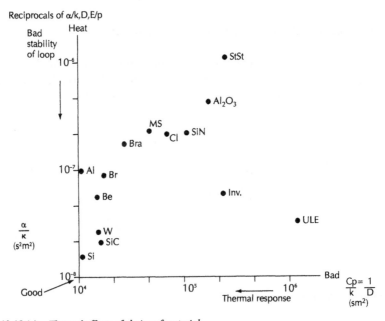

**Figure 13.18 (a)**    *Thermal effects of choice of material*

Notice that on all three graphs the silicon, silicon nitrides and carbides are best behaved. Conventional materials like mild steel, stainless steel, brass and bronze are poor. Hence, increasing use of semiconductor materials for their mechanical and thermal properties as well as their electronic properties makes them suitable for instrument design.

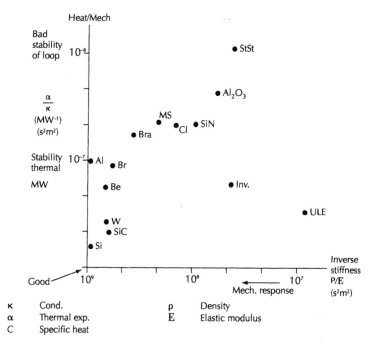

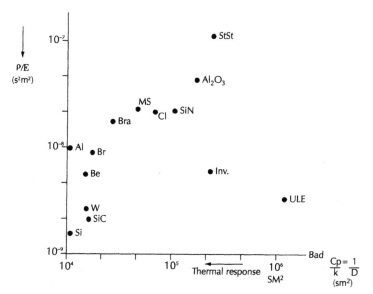

**Figure 13.18 (b)**   *Mechanical and thermal effects of choice of material*

**Figure 13.18 (c)**   *Mechanical vs thermal choice of material*

### 13.4.9   Mechanical stability

What appears at first sight to be an adequate design can, in fact, be greatly improved by changing the direction of forces – enlarging the triangle of forces and so on. Take as an example a roundness spindle (Figure 13.19 (a) and (b)). The leverage against vibration or noise in the top bearing is $\dfrac{b_1}{a_1}$, which can be large. Simply altering the design slightly by changing around the bearing mount (Figure 13.19 (b)) means that the $\dfrac{b_2}{a_1}$ ratio can be made small and stability is assured.

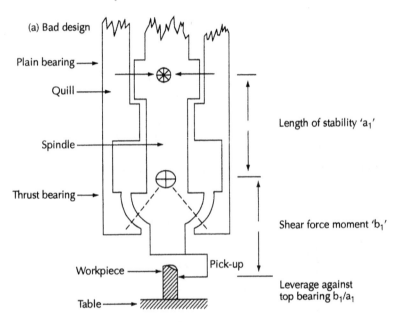

**Figure 13.19 (a)**   *Apparently well-designed spindle*

### 13.4.10   Optimized design

The early Talystep was intended to measure thin films and step heights to an incredible resolution that was far better than any instrument up until then. Reason came up with a technique that not only provided the correct degree of stiffness but also a traverse that enabled a widest range of specimen shapes. He used a very long ligament hinge 'S' for tremendous axial stiffness and yet used the weak lateral stiffness of the hinge to provide the stylus movement in the form of a curve. Hence, flat and curved specimens could be catered for (Figure 13.19 (c)).

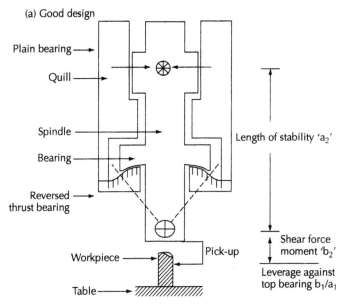

(a) Good design

Plain bearing

Quill

Spindle

Bearing

Reversed
thrust bearing

Length of stability 'a$_2$'

Shear force
moment 'b$_2$'

Leverage against
top bearing b$_1$/a$_1$

Workpiece

Pick-up

Table

**Figure 13.19 (b)**   *Actual well-designed spindle*

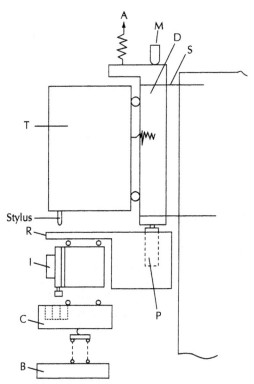

A

M

D

S

T

Stylus

R

I

C

P

B

**Figure 13.19 (c)**   *Talystep high-resolution measuring device*

### 13.4.11  Improvement in accuracy

The fact that it is only possible to improve accuracy on an already good instrument poses the question of how to determine the systematic or repeatable error of the instrument. Once found, it can be removed from all subsequent measurements. The general philosophy is illustrated in Figure 13.20.

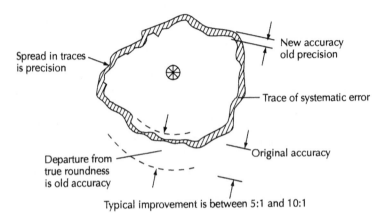

Figure 13.20  *Replacing the accuracy value of the spindle by its precision*

In what follows, the emphasis is on improving the performance of roundness instruments. The discussion and conclusions apply equally to straightness measuring instruments etc.

It has been pointed out earlier that a prerequisite of the instrument is that it has to be good enough to have a systematic error! Only under this condition can improvement be made.

### 13.4.12  Inverse method [13.8]

The Inverse or Reversal method shown in Figure 13.21 is for a roundness measuring instrument. A good specimen – usually the intended reference specimen – is measured with the roundness instrument in the normal way, then the instrument is stopped and the specimen turned through 180°. The pick-up is moved through the centre of rotation and the polarity of the transducer signal changed. Another measurement is then taken and the results analysed. Thus:

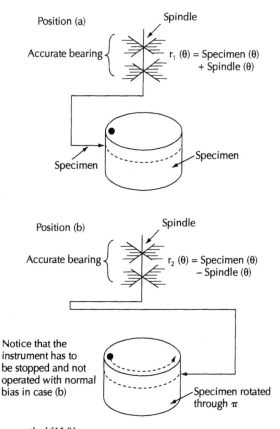

**Figure 13.21**   *Inverse method [13.8]*

$$\text{spindle error } (\theta) = \frac{r_1(\theta) + r_2(\theta)}{2}$$
$$\text{specimen } (\theta) = \frac{r_1(\theta) - r_2(\theta)}{2}$$

Providing that care has been taken when moving the specimen and that enough repetitions have been made to reduce the noise, then this method gives spindle and specimen values of deviation as a function of $(\theta)$. The calibrated specimen can be used in the future to check the spindle, providing that the specimen itself is stable and does not 'creep' with time.

Advantages of this method are:

1.   Fast result.

2.   Availability of unrestricted harmonics because it is a continuous signal with respect to $\theta$ that has been obtained.

Disadvantages are:

1.   The system has to be stopped.
2.   The pick-up has to move through the centre of rotation and have its polarity changed.

Exactly the same technique can be used for calibrating other instruments, for example, straightness (Figure 13.22).

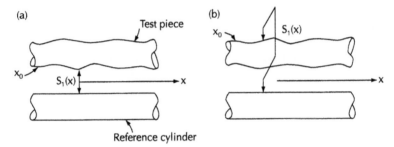

**Figure 13.22**   *Inverse method for straightness*

There is another method called **multistep.**

### 13.4.13   Multistep method [13.1]

In this method, the specimen is fixed to a rotatable table positioned under the spindle. The measurement of the specimen is made a few times. On the last run, the specimen is rotated through a fixed angle and the process repeated. After a few runs, the specimen is again rotated. This whole operation is repeated until the specimen is back in its original position (Figure 13.23.) A set-up is shown in Plate XIV.

The specimen errors and the spindle errors are unravelled from a matrix (Figure 13.24).

In this method, the spindle rotation is continuous. Also, there is no need to reverse the sensitivity polarity as in the inverse method.

Points about multistep method.
1.   Very accurate.
2.   Continuous measurement – the spindle is not stopped.

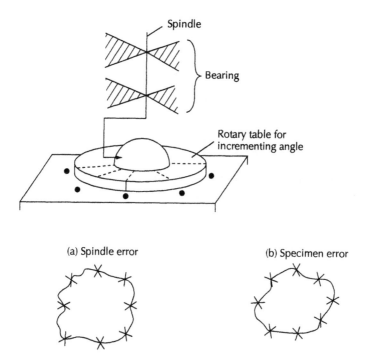

**Figure 13.23**   *Repeat or multistep method*

Readings α γ θ        θ is position (spindle)
                      α is position (specimen)

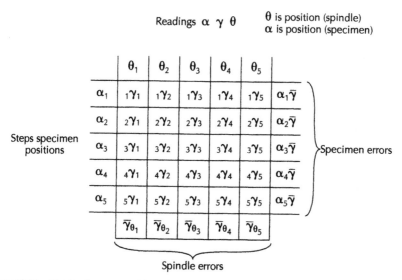

**Figure 13.24**   *Matrix of errors – multistep*

3.   Restricted harmonics.
4.   Slower in operation than the inverse method.

Note that with the multistep and inverse methods, some processing of the data is necessary but is not shown here for the sake of simplicity.

## 13.5   Instrument usage [13.9]

A number of elementary problems that keep on occurring are shown in Figure 13.25. The first ones show stylus effects.

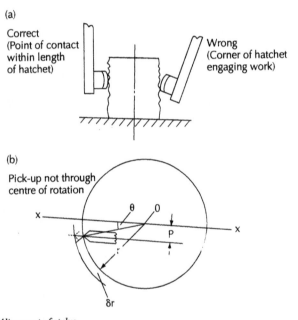

**Figure 13.25**   *Alignment of stylus*

With an offset probe that does not point through the centre of rotation of the instrument, a shift $P$ gives a small increase of magnitude of $\dfrac{\delta r \sec \theta}{\delta p}$ where $\theta = \sin^{-1} \dfrac{P}{r}$ [13.1]. The straightness of the probe column is very important because of the length of the probe.

If the probe is long, it magnifies any angular errors on the column with the result that the contact on the specimen is in the wrong place. This is an example of cosine error (Figure 13.26).

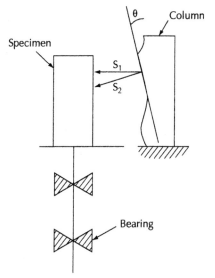

**Figure 13.26**   *Effect of lack of straightness of column*

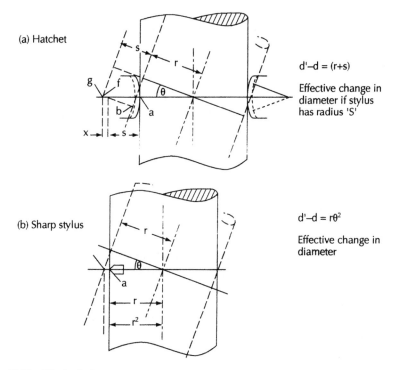

**Figure 13.27**   *Tilted cylinder*

The type of stylus is not too critical on ground shafts or smooth cylinders but if the cylinder is tilted, by presence of dirt on the base for example, the error in the signal can be significant as seen in Figure 13.27 – in proportion to the radius of the stylus S.

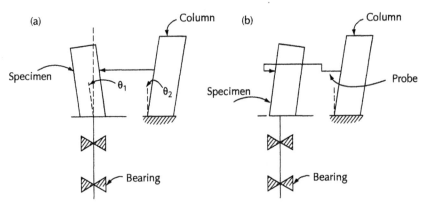

**Figure 13.28**  *Error inversion method*

The rotating table roundness instrument is not very suitable for detecting changes in the radius of, say, a cylinder as a function of height because if the probe column is tilted this will appear to be a taper on the specimen cylinder. However, it is possible to measure exactly what the tilt of the column is by using the inversion method (Figure 13.28) described above. This can either be used to help to adjust the column back to square or the tilt can simply be removed from the measured data.

## 13.6  Elastic design [13.11] – removal of hysteresis and friction errors

In previous sections, kinematics and mobility have been looked at. It has been noted that the limit of possible precision relies upon wear rate, hysteresis, lubrication/chemistry of the interface surface finish geometry and differences in the materials. One commonly used method to overcome these problems is to utilize the elastic distortion of a mechanism under an applied force. The simplest flexure mechanism is the hinge as shown in Figure 13.29.

The above design will produce small rotation with the rotation axis perpendicular to the plane of the paper. The axis of rotation in the above design is not clearly defined and may not even be stationary. The commonest method for obtaining a fixed rotation axis is to use a cross strip hinge.

The couplings shown above are commonly used in instrument mechanisms and can be commercially obtained over a wide range of sizes. However, for linear

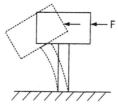

**Figure 13.29**  *Elastic deflection of hinge*

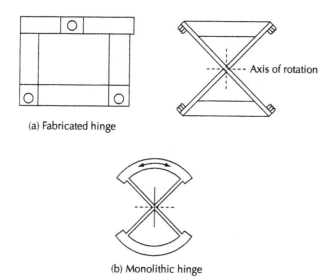

**Figure 13.30**  *Hinge rotation*

mechanisms, because of the diversity of applications it is usual to design these individually. Before looking at the design of these mechanisms, it is important to look at the advantages and disadvantages of elastic design.

### 13.6.1  The advantages and disadvantages of elastic design

1.   They are wear free. This means that the line of action of such devices (or the position of the axis of rotation) will remain constant.

2.   There are no rolling or sliding components. Surface finish effects and hysteresis due to contact friction will be eliminated.

3.   It is possible to manufacture these from a single (monolithic) piece of material. This eliminates problems associated with the high stresses necessary for the fixing of component parts or instabilities due to gluing. Fixing will also usually lead to creep.

4.   It is possible to construct mechanisms that are immune to temperature changes and even temperature gradients in certain planes.

5.   Displacements can be determined from a knowledge of the applied force.

6.   Failure mechanisms are usually catastrophic (i.e. fatigue or brittle fracture) and can thus be easily detected.

### 13.6.2   Disadvantages of elastic spring flexures

1.   The force required for a given displacement is dependent upon the elastic modulus.

2.   Hysteresis will exist due to the generation of dislocations in most materials. This is dependent upon the temperature and the type of bonding in the material.

3.   Flexures are restricted to small displacements.

4.   Out-of-plane stiffnesses are relatively low and thus great are must be taken to ensure that the drive axis is collinear with the desired motion.

5.   If high stresses are encountered or incorrect materials chosen, they may break, fatigue or work harden.

### 13.6.3   Linear spring mechanisms

The simplest linear spring motion is a cantilever beam (or a rod for angular rotation). Formulae relating the deflection to the applied force are readily available. The most commonly used linear flexure mechanism is the Jones type spring.

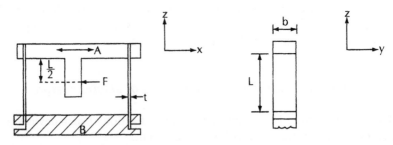

**Figure 13.31**   *Linear spring mechanism*

The stiffness ($\lambda_x$) in the x direction is given by:

$$\lambda_x = \frac{24EI}{L}$$

where $I = (\lambda t^3)/12$ second moment of area of each leaf.

It is apparent from Figure 13.31 that this mechanism will have a relatively low tortional stiffness about the z-axis. To increase this stiffness without altering any other values, each leaf can be split along the axes and the resultant two leaves separated (Figure 13.32).

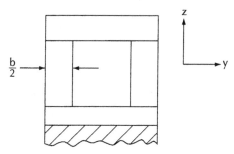

**Figure 13.32**   *Asymmetric hinge*

To prevent bending about the y-axis, the drive force should be applied at the mid-point between the two platforms so long as there are no significant loads on the platform. There are two inherent errors associated with this design.

a)      Upon translation, platform A will tend towards platform B due to the curvature of the springs. This is a difficult problem to solve mathematically and is only negligible for small displacements.

b)      Thermal expansion effects will directly affect the position of A relative to B. The change in position of A relative to B for a given temperature change $\Delta T$ is given by $\delta \alpha \Delta T$. This effect is obviated by the use of a compound rectilinear spring (Figure 13.33).

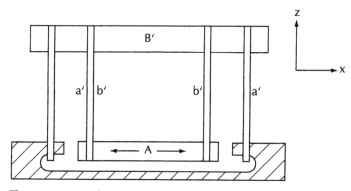

**Figure 13.33**   *Floating asymmetric hinge-position of specimen table A*

In this device, a force F is applied to platform A. This will result in a displacement relative to B′ of magnitude $F/\lambda_b'$ between B′ and the base B. If all the leaf springs are identical, then the total displacement $\delta$ is given by $\dfrac{2F}{\lambda}$ where $\lambda$ is the composite stiffness.

Thus, it can be seen that the stiffness of the mechanism is halved by the addition of an identical spring in series. However, upon a finite temperature change, the thermal expansion of the leaf support springs a′ and b′ will be equal and opposite. This will result in zero net motion of platform a relative to the base. This system will also be immune to temperature gradients in the y-axis.

The analysis in Section 13.4.5 showed that this mechanism has two degrees of freedom. This can be reduced by using an even more symmetrical design of the form (Figure 13.34).

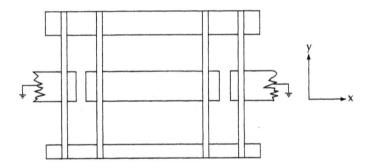

**Figure 13.34**   *Linear symmetrical movement*

If each joint is considered to be a single degree of freedom hinge, then the mobility of this mechanism is given by:

$$M = 3(12-1) - 2(16)$$

From this analysis, this mechanism is a single degree of freedom system such that it is constrained to move along a singular path. It would be expected that the application of a force to this platform will result in a rectilinear motion.

There are still problems associated with this particular arrangement. That is, the device is fabricated from many components and, although machining this shape from a monolithic block is feasible, the accurate machining of thin flexures will be both difficult and expensive.

To overcome this problem, it is common to use a necked hinge instead of the leaf spring. These hinges can then be combined to form any of the aforementioned devices. An example of a compound monolithic rectilinear spring is shown in Figure 13.35.

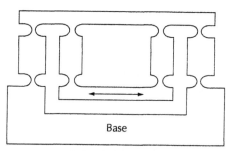

**Figure 13.35**   *Simple construction method for springs*

## 13.7   Workpiece cleaning and handling

### 13.7.1 Workpiece cleaning [13.11]

From the point of view of engineering metrology, there is usually not much time to prepare a specimen workpiece prior to measurement. All that can be attempted in in-process measurement is a contact wiper on the surface or a brush for internal holes. Sometimes, compressed air is used to clear the surface of debris and oil films. Some early attempts to measure surfaces during operation had the compressed air in the instrument, thereby preventing dirt from settling on any optical or electronic components, especially those elements near to the surface.

The situation in the case of in situ measurement when the specimen is removed from the machine depends on the application. In the case of bearings, for example, it is essential to clean the raceway or ball of grease and oil. Once removed from the machine, a means for turning or moving the specimen is needed that requires a fiction drive roller somewhere on the periphery of the race. If the part is oily, slippage occurs and the effective traversing speed becomes intermittent. It does not matter if the rotation is encoded but if not then filtering and parameter assessment is difficult.

A cleaning agent such as acetone is needed to clean the part before measuring.

### 13.7.2   Handling [13.10]

Moving the work to the instrument entails touching the work. Some people, unfortunately, have acid sweat, which marks and often ruins the surface with rust if the workpiece is steel or iron. This is not apparent immediately but repeated use of comparison specimens can be disastrous. People who have 'rusty fingers' have to wear gloves.

## 13.8   Conclusions

Some aspects of the issues involved in the design and identification of error in instrument design has been given. These were kinematics, mobility, materials, error reduction and elastic effects.

The overall trend is for the instrument and components to become smaller and to be able to measure dimensions as well as texture.

Points to note.

1.   Metrology loops should avoid force loops.
2.   All forces should be kept to a minimum.
3.   Any vibration or heat sources should be eliminated near to their origin.
4.   Loops should be as small as possible.
5.   Any unavoidable detrimental input should be applied symmetrically to the loop.
6.   Errors should be removed by inversion or multistep if movement errors are repeatable.
7.   Hysteresis errors should be reduced by elastic design.

It is important to design instruments in a way that best reflects what the measurement is for. In manufacture, in using the surface to monitor the machine tool for example, the probe of the instrument should ideally mimic the tool path if commercially possible. Failure to do this results in difficulties. For example, a milled surface is very difficult to characterize using a conventional instrument yet the tool paths are known and are straightforward. Bringing the use and measurement together is one future possibility for engineering metrology.

## References

13.1   Whitehouse D. J. *Error Separation Techniques in Surface Metrology*. Pro. Inst. Phys. J. Sci. Inst. Vol. 9 p361 (1976).

13.2   Smith S. T. and Chetwynd D. G. *Foundations of Ultra Precision Mechanism Design*. Gordon and Breach UK (1992).

13.3   Moore W. R. *Foundations of Mechanical Accuracy*. Moore Special Tools Bridgeport Conn. (1970).

13.4   Pollard A. F. C. *The Kinematic Design of Couplings in Instrument Design*. London Hilger and Watts (1929).

13.5   Fichter E. F. *A Stewart Platform Based Manipulation* . Int. Journal of Robotic Research Vol. 3 No. 2 p157 (1986).

13.6   Huang T. Wang J. and Whitehouse D. J. *Closed Form solution to the Position Workspace of Stewart Parallel Manipulators*. Science in China Series Vol. 41 No. 4 p393 (1998).

13.7   Chetwynd D. G. *Structural Materials for Precision Devices*. Pre. Eng. Vol. 19 No. 1 p3.

13.8   Donaldson R. R. *A Simple Method of Separating Spindle Error from Test Ball Roundness*. CIRP annals Vol. 21 p125 (1972).

13.9   Reason R. E. *Report on the Measurement of Roundness*. Rank Precision Ltd. (1966).

13.10  Bennett J. M. and Mattsson L. *Roughness and Scattering* Ch. 8. Optical Soc. of America Washington D. C. (1989).

13.11  Jones R. V. *Some uses of Elasticity in Instrument Design*. J.Sci. Instr. Vol. 39 p193 (1962).

# 14

# Calibration of instruments

## 14.1 General

Calibration of roughness and roundness instruments is difficult because of the small size and special nature of the features being measured. The difficulties have been recognized for some time and have led to an extensive use of carefully prepared standard specimens. These will be outlined in Section 14.1 and then given in more detail later in the text in addition to some specialized procedures.

### 14.1.1 Calibration specimens for roughness

There are a number of standards concerning this area, including ISO 5436 for documentation.

| | |
|---|---|
| ISO 468 | Roughness – parameters. |
| ISO 1878 | Classification of instruments and devices for the measurement and evaluation of the geometrical parameters of surface finish. |
| ISO 1879 | Instruments for the measurement of surface roughness by the profile method – vocabulary. |
| ISO 1880 | As above – contact (stylus) instruments of progressions profile. Transformation – profile recording instruments. |
| ISO 3274 | As above – contact profile meters, system M. |

## 14.1.2   Types of calibration specimens [14.1]

To cover the range of requirements, four types of specimens are described. These are:

*Type A* – for checking vertical magnification

**A1**  Wide calibrated groove with flat bottom, or a number of separated grooves of equal or increasing depth. The groove has to be wide enough to be insensitive to the condition of the stylus.

**A2**  As A1 but with rounded bottom.

*Type B* – for checking the stylus tip

**B1**  Number of separated grooves proportioned to be increasingly sensitive to the dimension of the stylus. These are for instruments sensitive to displacement – the usual sort.

**B2**  These have two grids of nominally equal $R_a$, one being sensitive and the other insensitive to the dimensions of the stylus tip. They are used for velocity sensitive pick-up e.g. piezoelectric, the ratio of the $R_a$ values from both grids being taken as the criterion for wear.

*Type C*

These are used for checking parameters. They comprise of a grid of repetitive grooves of simple shape, which have relatively low harmonic content. These are triangular, sinusoidal etc. They can be used for checking horizontal magnification if the groove spacing is accurate.

Type C specimens of differing waveforms have to be compatible.

The declared parameter values issued with the standard refer to instruments having a smooth straight datum (not skid) and filtered profiles obtained from the trace according to ISO 3274. The parameter values have to be declared with reference to the stylus tip. For skid type instruments, the skid should not penetrate significantly. This is usually achieved by having groove spacing to as close as possible that permitted by the stylus.

*Type D*

These specimens are for overall checks of meter calibration. Specimens should have irregular profiles as in grinding. They should have constant cross section in depth. They simulate workpieces containing a wide range of crest spacings.

In more detail, these are as follows:

*Type A*

| Nominal values for type A1(μm) | | | | | |
|---|---|---|---|---|---|
| Depth | 0.3 | 1.0 | 3.0 | 10 | 30 | 100 |
| Width | 100 | 100 | 200 | 200 | 500 | 500 |

| Nominal values for type A2 | | | | |
|---|---|---|---|---|
| Depth (μm) | 1.0 | 3.0 | 10.0 | 20.0 | 100 |
| Radius (μm) | 1.5 | 1.5 | 1.5 | 0.75 | 0.75 |

*Type B*

B2: two grids on common base.

(i)  Sensitive grid

$\alpha = 150°$
$R_a = 0.5$ μm $\pm 5\%$ $S_m$ 15μm

(ii)  Insensitive grid

$R_a = 0.5$ μm $\pm 5\%$ $S_m = 0.25$mm

*Type C*

Sine wave    Type C1 Nominal values $R_q(\mu m)$

| $S_m$ | (mm) 0.08 | 0.25 | 0.8 | 2.5 |
|---|---|---|---|---|
| | | | | |
| $R_a$ | 0.1 | 0.3 | 1 | 3 |
| | 0.3 | 1 | 3 | 10 |
| | 1 | 3 | 10 | 20 |
| | 3 | 10 | 30 | – |

*Type D*

These are irregular ground specimens repeated every 4 mm.

Nominal filtered values $R_q$ μm.

| Nominal $R_q$ | Tolerance | Uncertainty of measurement % | Standard deviation from meas. % |
|---|---|---|---|
| 0.15 | ±30 | 5 | 4 |
| 0.5 | ±20 | 3 | 3 |
| 1.5 | ±15 | 3 | 3 |

For other details such as marking, materials etc. see ISO 5436 1985 (E).

The starting point has to be the measurement of the stylus of the instrument.

## 14.2   The stylus calibration

As mentioned in Section 6, there are a number of styli in use. One is the 90° diamond pyramid and another the 60° diamond cone as shown in Figure 14.1.

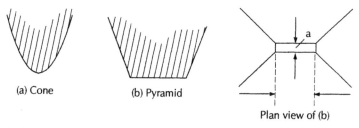

(a) Cone          (b) Pyramid          Plan view of (b)

**Figure 14.1**   *Diamond shapes*

It is possible to measure these styli with an SEM (scanning electron microscope) providing the stylus does not charge up. If this happens, then some sort of conductor has to be deposited e.g. gold. Unfortunately, such a procedure is hardly viable in an industrial workshop, so the instrument manufacturers have devised a block standard that can be used relatively easily. One of these is shown in Figure 14.2. This corresponds to a B1 standard.

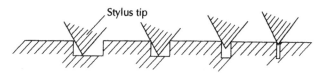
Stylus tip

**Figure 14.2**   *B1 block standard*

The block is usually quartz and coated with chrome or similar element. A number of grooves are ruled in the block, having different widths. There are usually four or five grooves but there is no limit.

The unknown stylus is tracked across the blocks as seen in Figure 14.2. In the very wide groove, the stylus touches the bottom, whereas where the grooves get progressively narrower, the stylus finds it increasingly difficult to get to the bottom. There comes a point when it does not bottom and then a further point where it hardly registers at all. When this stage has been reached, all that remains is to record these readings of groove penetration on a graph (Figure 14.3) of groove width versus penetration.

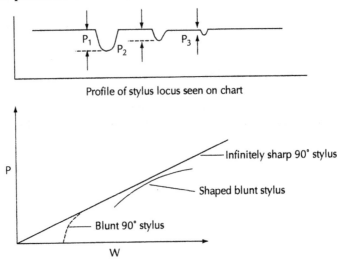

*Profile of stylus locus seen on chart*

**Figure 14.3**   *Stylus tip measurement*

From the graph, it is possible to work out what the stylus tip dimension is and also some idea of the shape of the tip. This technique has its problems because the grooves get filled with debris and misleading results occur.

It is more comprehensive than one of the original stylus tests, the caliblock, introduced by General Motors in the 1940s. This was simply a block with two ruled patches on it. The rulings were triangular (Figure 14.4).

**Figure 14.4**   *The caliblock*

The one block was 125μ" $R_a$ and the other 25μ" $R_a$. The height was calibrated against the rough standard for the low magnifications. The fine patch was used for the higher magnification. These were measured off the chart. Also, the $R_a$ value was calibrated and put on the fine patch. The practical $R_a$ value obtained was compared on a chart of the decrease in $R_a$ with stylus size (worked out previously). This gave a good idea of the stylus tip dimension. Unfortunately, neither of these methods could give an effective warning of a stylus that had been chipped and had a sharp point (Figure 14.5).

**Figure 14.5**   *Chipped stylus*

More exotic methods have been tried to deduce the stylus tip dimension from a series of measurements on known specimens using a neural network scheme. This idea proved to be disappointing because the network has to be trained on a number of styli and specimens and it soon emerges that any stylus that is slightly out of the ordinary gets the whole scheme confused.

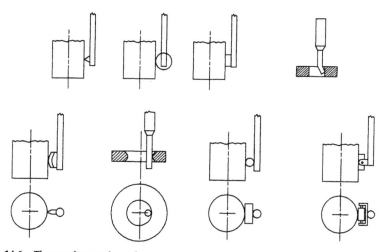

**Figure 14.6**   *The roundness stylus – shape*

The number of styli available in roundness and similar measurements is very large indeed. The nature of roundness is such that it does not require a sharp tip. Various styli are shown above in Figure 14.6 in side and plan view.

## 14.3   Height calibration

### 14.3.1   Type A1 standards

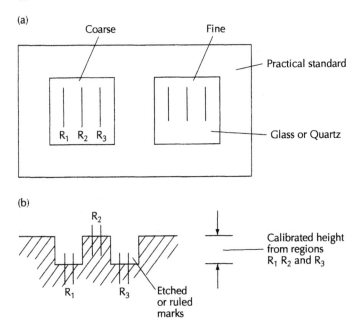

**Figure 14.7**   *Height calibration*

The standard described above has been used for many years and is shown in Figure 14.7. This again had two patches on it. As before, it was usually made of quartz with chrome on it. Each patch had three grooves ruled in it. The top of $R_2$ and the valleys of $R_1$ and $R_3$ are used in the calibration procedure.

Notice the fact that it is only the mid third of these levels that is used as the standard height. This is intentional because it is necessary to take account of some wear and edge break-up. The working regions are shown with parallel lines in Figure 14.7.

Although the use of gauge blocks was considered for use in height calibration of surface instruments, it soon became clear that even the closest slip gauges were too coarse for some applications. Reason used a lever system to scale down the gauge height difference [14.2]. He contrived a system in which the pivot position was calibrated out in an ingenious way. The problem was that even when using a 20:1 reduction lever, the step was still too big. It was the advent of instruments purporting to measure down to nanometres and below that made a rethink necessary. Basically, one rule of metrology is being broken when using conventional ruled and etched standards. This rule is that the unit of measurement should be within an order of magnitude in size of the detail being measured. Even the wavelength of light was too large by a factor of about a thousand. Clever interpolation methods enabled measurements in the nanometre range to be carried out, but this method is straining the technology.

### 14.3.2  Crystallographic method

The answer to the calibration problem is to use a natural dimension of the order of nanometres. Crystal lattice spacing is an obvious way forward. Topaz as a crystal produces fine steps when cleaved but the steps can be slightly ambiguous. What is required is a crystal that has only one cleavage plane that produces a sub-nanometre step. Unfortunately, it has to be hard (~ 7 or 8 on the scratch test scale) and preferably conducting [14.3].

The method which is now being used to calibrate the most sensitive instruments uses the crystal lattice spacing of silicon as the reference unit of measurement (Figure 14.8) [14.4, 14.5].

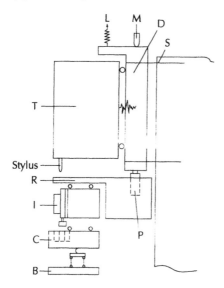

**Figure 14.8**  *Step height instrument*

Instead of the direct method using dislocations and the steps mentioned above, it uses an indirect method via X-rays. The method in effect uses an X-ray interferometer to enable the lattice spacing to be utilized (Figure 14.9 overleaf).

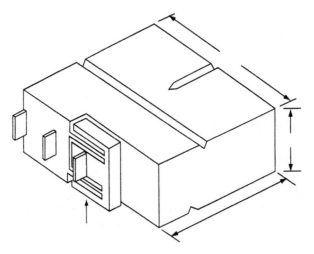

**Figure 14.9**  *X-ray interferometer*

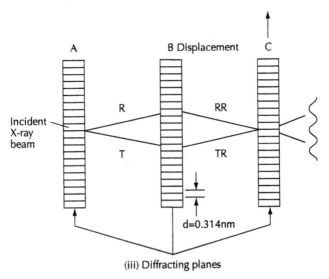

**Figure 14.10**  *X-ray path through silicon interferometer*

The interferometer is a monolith crystal of silicon. This has three blades A B C, as in Figure 14.10, milled out of the block. The blades A and B are fixed and the third, C, is capable of a small vertical movement by virtue of the fact that it is a small ligament hinge. The X-rays enter from the left and diffract at A to produce two X-ray beams, which hit B and then diffract. The two inner beams focus on C. The probe of the instrument being calibrated rests on top of blade C.

A mechanism under C pushes the blade upwards and the output of the probe on top of C is recorded. At the same time, the X-rays passing through C are alternately in and out of phase with the crystal lattice. This produces a sinusoidal variation in X-ray intensity at the detector. The wavelength is the lattice spacing. Two separate outputs are recorded from the apparatus; one from the probe of the instrument under test and one from the X-ray interferometer. So, the probe is calibrated relative to the crystal lattice spacing of silicon, which is 0.3μm. As the X-ray output is sinusoidal, it can be interpolated by orders of magnitude to give a potential height calibration of fractions of atoms.

## 14.4 General height

For a wide range surface instrument, a standard has been devised that is no more than a hemisphere of known radius, sometimes of 80mm radius (Plate XVII) or lower (12.7mm or 22mm). The standard hemisphere is made of tungsten carbide. The various size balls are necessary because of the different stylus arms. Typically, the 12.7mm ball is used with a standard 60mm stylus arm, so it is possible to calibrate instruments from the relatively coarse to the extremely fine [14.6].

The standards above are for texture and form instruments. Other instruments such as roundness instruments can also be calibrated using a polished hemisphere, as will be described next.

## 14.5 Roundness calibration – for height

Two common ways of calibrating roundness instruments are shown in Figure 14.11 (a) and (b). The (a) method is simply a flat milled out of an accurate cylinder. When

(a)

(b)

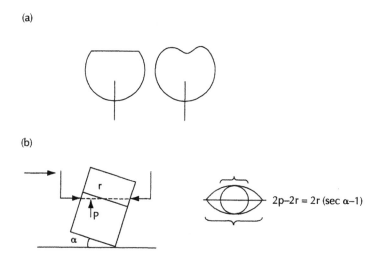

$$2p-2r = 2r\,(\sec \alpha - 1)$$

**Figure 14.11** *Calibration of roundness instruments (a) machined flat on cylinder and (b) tilted cylinder*

(a)

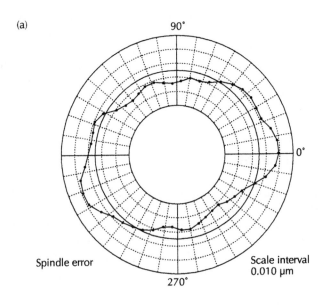

90°

0°

Spindle error

Scale interval
0.010 µm

270°

(b)

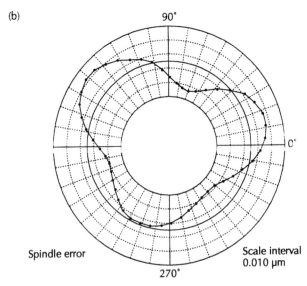

90°

C°

Spindle error

Scale interval
0.010 µm

270°

**Figure 14.12**   *(a) and (b) roundness calibration*

magnified to the level suitable for atypical roundness, the flat looks extremely
concave as can be seen in the figure. This method is simple and quite accurate to 1%.
Method (b) is even simpler. It is just a tilted cylinder, which is measured with a sharp
stylus. The output is an ellipse, which can be measured easily. Perhaps the main disadvantage of this method is the fact that if debris gets stuck under the cylinder, then
the fixed angle $\theta$ might be affected and so give the wrong answer.

The accurate way of calibrating roundness has been described in Chapter 13 using the inversion method and the multistep method. The multistep calibrates a reference sphere, which can be used as a reference standard for the instrument. A graph is shown in Figure 14.12 (a) giving a plot of a typical fine calibration hemisphere. Figure 14.12 (b) shows the plot for the spindle of the instrument. These results were taken using the system shown in Plate XVIII(a) and from [14.5].

## 14.6   Filter calibration (type C standard)

Ideally, what is required is a standard that calibrates from the stylus through to the recording device. Many types of ruled standard have been tried, the ruling spacing being a simulation of a sine wave through the internal filter. Figure 14.13 illustrates some examples. One problem immediately obvious is the sharp edge in some of the standards. These edges cause problems when traversing. The stylus has to be prevented from dropping off the end of the specimen.

From the figure, it is seen that there are a number of possible shapes that can be used. Probably the easiest is the triangular shape, which is obtained by a ruling machine with a triangular tool. The sinusoidal standard can be made but has some problems.

Everyone thinks that a sinusoidal shape is perfect but it should be remembered that most surfaces are random and require a random standard to be used for calibration. These are type D standards. See Figure 14.15 [14.4].

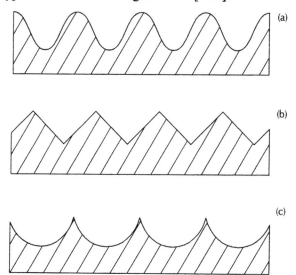

**Figure 14.13**   *Filter standard*

Obviously, the same standard can be used as a height standard and a length standard if the specimen has been calibrated properly.

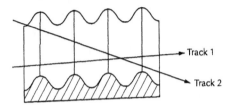

**Figure 14.14**   *Inclined filter standard*

An ingenious way of extending the range of filter standards is shown in Figure 14.14. To get longer wavelengths, the trace is taken at an angle to the normal direction. This type of approach is limited but a range of 3:1 in wavelength is possible. There are some slight problems caused by the stylus not riding over the ridges at 90° but generally these effects are negligible.

Random waveform standards are needed to cover a wide range of wavelengths. One of these has been made by PTB (Hillman) in Germany [14.7]. The highly successful process used to make this standard is plunge or creep feed grinding (Figure 14.15).

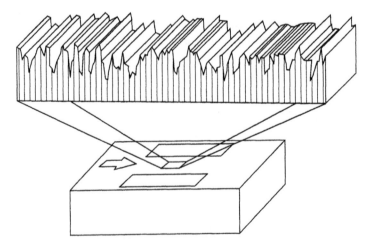

**Figure 14.15**   *Type D standards*

The standards mentioned above are conventional. Some newer standards are being developed, which could in the near future supersede the old ones. (see Plate XIX).

Figure 14.16 shows the problem found when tactile roughness standards are used to calibrate or try to calibrate optical instruments. In (a), there is a profile of a

(a)

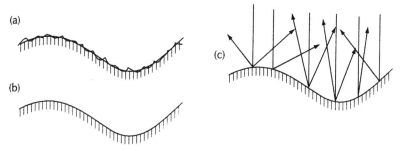

(b)

(c)

**Figure 14.16**   *Suitability of stylus standards for optical use*

sinusoidal standard that has been diamond turned. Underneath (a) is profile (b), which is the result obtained using a stylus instrument. Notice that the turning marks are effectively smoothed out by the integrating effect of the stylus spoken about earlier. Profile (c) is the result of using the standard as an optical standard. The light, in effect, differentiates the standard. All that is seen is the scatter from the turned marks. The actual sinusoidal shape of the standard is almost all lost. There is no significant correlation between the optical result and the stylus result.

## 14.7   Vibrating tables [14.8]

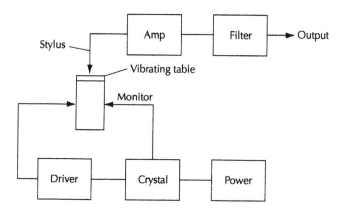

**Figure 14.17**   *Vibrating tables*

Instead of making ruled or photo-etched standards for checking the filter, one method that is being used is that of a vibrating table as seen in Figure 14.17. The stylus is placed on the table, which is then made to vibrate at different frequencies. The table is monitored to control the frequency and amplitude of the vibration. The output of the instrument is then observed. It is quite easy to get a set of transmission curves calibrating the system: its versatility is attractive. The only problem with this is that the effects of tracking across a specimen are lost because the stylus does not move laterally and the result may be somewhat artificial.

## 14.8   Instrument comparisons

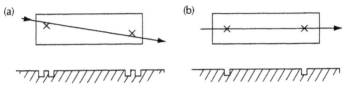

**Figure 14.18**   *Comparison of results using different instruments*

There have been, and still are, many comparisons between instruments being carried out. Often, one or more specimens are circulated around the world to the principal laboratories and the results compared.

It is astonishing to see the divergence of results. Many laboratories take many measurements so that the coverage value is stable but still differences occur. A much neater way to compare results is shown in Figure 14.18. Only one trace is required in this method.

Two crosses are made in the specimen under test, then the instrument is made to trace the specimen until the probe intercepts the crosses points as seen in Figure 14.18 (b). In the comparison, all that is required is that every trace from every laboratory traverses through just these crossing points.

This method is simple and cheap. It is quite sufficient to make the cross points with a razor. Using this simple trick reduces errors dramatically.

### 14.8.1   Calibration of scanning instruments

There is a trend in scanning instruments, in particular scanning electron microscopes, scanning tunnelling microscopes etc. to only partly calibrate the instrument.

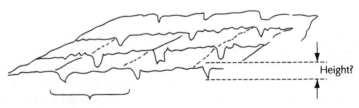

**Figure 14.19**   *Result comparison*

In most cases, the lateral structure is calibrated but not the height (Figure 14.19). This used not to be important but today it is vital to calibrate in both z and xy directions. In many instrument makers, the calibration of height – if any – is only attempted internally for control purposes.

What is happening is that it is only the scanner that is being calibrated and not the transducer! Stylus scanners do not have this problem. Neither would a step height resulting from a lattice jump on a crystal.

## 14.9    Chain of standards

The following tables are set out according to ISO standards and are meant to give a comprehensive breakdown of all aspects of a measured parameter. Table 14.1 shows the chain of standards for roundness; usually up to six links are in the chain. All the chain links have to be acted upon otherwise the procedure is invalid.

Roundness, straightness and roughness data are given below.

**Table 14.1**    *The out-of-roundness chain of standards*

| Chain link number | | 1 | 2 | 3 |
|---|---|---|---|---|
| Geometrical characteristic of feature | Parameters | Codification on a drawing | Definition of tolerance | Definitions for actual feature |
| Form of line independent of datum | Roundness | 1101, 2768-2 | 1101, 2768-2,6318 | 1101 |

| Chain link number | | 4 | 5 | 6 |
|---|---|---|---|---|
| Geometrical characteristic of feature | Parameters | Comparison with tolerance limits | Measurement equipment requirements | Calibration requirements |
| Form of line independent of datum | Roundness | (5460) | 4291, 4292, 463 13365, 10360 | 1101 |

**Table 14.2**   *Chains of standards*

| Chain link number | | 1 | 2 | 2 |
|---|---|---|---|---|
| Geometrical characteristic of feature | Parameters | Codification on a drawing | Definition of tolerance | Definitions for actual feature |
| Form of line independent of datum | Straightness | | | |

| Chain link number | | 4 | 5 | 6 |
|---|---|---|---|---|
| Geometrical characteristic of feature | Parameters | Comparison with tolerance limits | Measurements equipment requirements | Calibration requirements |
| Form of line independent of datum | Straightness | | | |

**Table 14.3**   *The surface roughness chains of standards*

| Chain link number | | 1 | 2 | 3 |
|---|---|---|---|---|
| Geometrical characteristic of feature | Parameters | Codification on a drawing | Definition of tolerance | Definitions for actual feature |
| Surface roughness | M-system $R_a$ | 1302 | 4287-1,-2, 468 | 4288 |
| | M-system other | | 4287-1, 468 | |
| | Motif method R | | | 12086 |

| Chain link number | | 4 | 5 | 6 |
|---|---|---|---|---|
| Geometrical characteristic of feature | Parameters | Comparison with tolerance limits | Measurements equipment requirements | Calibration requirements |
| Surface roughness | M-system $R_a$ | 4288, 2632-1,-2 | 3274, 1878, 1879, 1880, 2632, 11562 | 5436, 2632 |
| | M-system other | | 3274, 1880, 11562 | |
| | Motif method R | | | 12086 |

**Table 14.4**   *The form of surface – independent of datum – chain of standards*

| Chain link number | | 1 | 2 | 3 |
|---|---|---|---|---|
| Geometrical characteristic of feature | Parameters | Codification on a drawing | Definition of tolerance | Definitions for actual feature |
| Form of a surface independent of datum | Cylindricity | 1101, 2768-2 | 1101, 2768-2 | 1101 |
| | Flatness | 1101, 2768-2 | 1101, 2768-2 | 1101 |

| Chain link number | | 4 | 5 | 6 |
|---|---|---|---|---|
| Geometrical characteristic of feature | Parameters | Comparison with tolerance limits | Definition of tolerance | Definitions for actual feature |
| Form of a surface independent of datum | Cylindricity | (5460) | 10360-1, -2, -3, -4 | |
| | Flatness | (5460) | 463, 9493, 8512, 10360-1, -2, -3, -4 | |

## 14.10   Conclusions

Calibration of instruments should, where possible, be carried out on the instrument in working mode. The standard should take this into account. For example, height standards should be capable of being used effectively when the instrument is traversing.

An artefact such as a height standard should be calibrated in a medium that is close to the function of the surfaces to be subsequently used. For example, if the surfaces are to be used in a tribological regime, then a stylus method should be used for calibrating the standard and measuring the workpiece to be used. Mixing media e.g. calibration optically and using in a contact situation breeds errors.

The unit of measurement should be as close in size as possible to the features of importance in the function of the workpiece. If the application is on the atomic scale, the unit should be atomic (or at least be within a hundredth of it). In this case, crystal lattice spacing is a solution. Using optics, for example, in this case although possible, places a strain on techniques such as interpolation.

# References

14.1   ISO 5436. *Calibration Specimens.* p784 ISO Handbook 11 (1999).

14.2   Reason R. E. *The Measurement of Surface Texture.* Ed. Wright Baker Longman London (1971).

14.3   Whitehouse D. J. '*Stylus Methods' Characterization of Solid Surf*aces. Kane and Larobee. New York, Plenum Ch. 3 (1973).

14.4   Chetwynd D. G. Siddon D. P. and Bowen D. K. *X ray Interferometer Calibration of Micro Displacements Transducer.* J. Phys. E, Sci. Inst. Vol. 16 p871 (1988).

14.5   Whitehouse D. J., Bowen D. K., Chetwynd D. G. and Davies S. T. J. Phys. E Sci. Inst. Vol. 21 p40 (1988).

14.6   Taylor Hobson. Centre of Excellence.

14.7   Hillman W. Kranz O. and Eckol T. Wear Vol. 97 p27 (1984).

14.8   Thwaites E. G. Int. Symp. Pm Metrology Conf. Proc. INSYMET 74 Bratislava (1974).

# Sampling, numerical analysis, display

## 15.1 General

It is impossible to cover such a field in a few lines yet it has to be mentioned in order to draw attention to the problems.

The development of instruments up to the mid 1960s was relatively straightforward. Practically all aspects of the instruments were analogue. One aim of these instruments was to provide analogue circuitry, which when acted on the roughness data, provided a measure according to a specified parameter. For example, to simulate $R_a$, the mean line was established by an analogue circuit. In the early days, this was a 2CR filter. The actual data was subtracted continuously from this mean with this line and the rectified result averaged over the assessment length to give $R_a$. In all this, the investigator had no access to the raw data emanating from the pick-up and amplifier. What the investigator saw was either a meter reading or a chart. The integrity of the signal was maintained from pick-up to chart. In fact, the frequency response of the recorder was matched to the frequency equivalent of the stylus tip at the tracking speed. The fact that there were many approximations made in terms of filter and parameter circuitry did not seriously jeopardize the resultant roughness value. The investigator simply used what was made available by the instrument maker: the user had no say about the way in which parameters were measured. This all changed with the advent of digital in-surface metrology methods in 1965 [15.1]. Earlier examples [15.2] only dealt with data collection and display.

It soon became relatively easy to tap off the raw data from the pick-up amplifier. Once outside the original instrument, the responsibility of the instrument maker to present sensible values of the surface parameter and chart was finished. The matching of each unit of the instrument in terms of frequency response and noise level no longer exists.

Unfortunately, many investigators had different ideas about parameters. Some knew little about numerical analysis. This produced the 'parameter rash' [15.3].

Given good analysis, the freedom allowed by digital methods is very useful. It became possible in theory for the customer to match a surface parameter to his or her requirements. Practically, this situation was rarely achieved because of the lack of time or effort on behalf of the user or researcher to master the new digital discipline. In what follows, a few examples will be given.

## 15.2   Numerical model

Probably the biggest problem confronting instrument makers and users is the lack of conformity of all aspects of digital analysis ranging from software to sampling. Consider, for example, numerical analysis. There are two aspects. One is concerned with numerical accuracy.

Take curvature, for example. Many users would identify this with the second differential $\dfrac{d^2y}{dx^2}$ and the numerical differential as:

$$\frac{d^2y}{dx^2} = -\left(\frac{2y_0 - y_1 - y_{-1}}{h^2}\right)$$

where the data points are shown as in Figure 15.1.

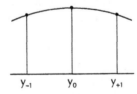

$$y_{-1} \qquad y_0 \qquad y_{+1}$$

**Figure 15.1**   *Three point model*

Some investigators would consider the three-point model shown to be adequate. Unfortunately, it is a crude approximation to the true answer.

A nearer approximation is a Lagrangian extension to the formula above.

$$C = -\frac{(2y_3 - 27y_2 + 270y_1 - 490y_0 + 270y_{-1} - 27y_{-2} + 2y_3)}{180h^2\left(1 + \dfrac{1}{60h}(y_1 - 9y_2 + 45y_1 - 45_{y-1} + 9_{y-z})^{\frac{3}{2}}\right)}$$

This is considerably sounder to use, even if slightly more complicated.

## 15.3   Sampling criteria

Table 15.1 gives the recommended sampling rate for the Gaussian filter used in roundness. According to the table, it is necessary to take five sample values of roundness in order to capture all detail. This is considerably higher than the well-known Nyquist criterion, which claims that 2 samples per undulation per rev should suffice.

The reason for this apparently excessive sampling rate is shown in Figure 15.2.

**Table 15.1**   *Roundness ISO/12 181 –1-2-3 Gaussian filter*

| Nominal part diameter (mm) | UPR | Minimum number of sampling points |
|---|---|---|
| D ≤ 8 | 15 | 75 |
| 8 < D ≤ 25 | 50 | 250 |
| 25 < D ≤ 80 | 150 | 750 |
| 80 < D ≤ 250 | 500 | 2500 |
| 250 < D | 1500 | 7500 |

Equivalent to $\lambda_s = 0.8$mm

Stylus sizes : 0.015; 0.05; 0.15; 0.5; 1.5; 5; 15 and 50mm

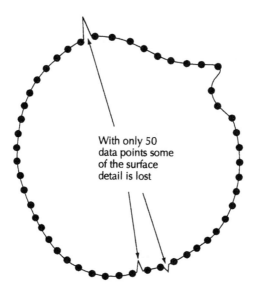

With only 50 data points some of the surface detail is lost

**Figure 15.2**   *Missing crucial peak*

It is clear from Figure 15.2 that even with many samples taken by roundness instruments, the odd freak peak can still be missed and would consequently affect the calculation of centre position. The situation is very much worse when measuring a hole. Because so few sample points are taken, the centre position can vary considerably in position – depending on the individual set of data points. Notice in Figure 15.3 that the first picture (a) shows the centre found by a tapered plug gauge pushed into the hole: a case when the old fashioned method is more accurate than the modern. Figures (b) and (c) show that if only four points are used, the error in the centre can be considerable.

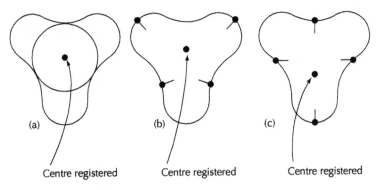

**Figure 15.3**  *Different centres for same geometry*

One unusual problem that has arisen recently and which is somewhat disconcerting is the fact that in some cases, the accuracy of measurement is worsened as digital methods are introduced as already seen in Figure 15.3. Also, different methods use different number of points. The roundness instrument has many sample points whereas the CMM has four or thereabouts (see Figure 15.4).

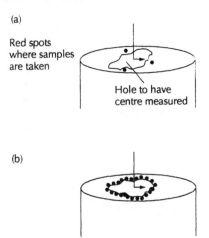

**Figure 15.4**  *Too few and too many points*

Figure 15.4 (a) shows too few points taken – typical of a CMM, whereas Figure 15.4 (b) shows too many points taken, perhaps by a roundness instrument. Both are wrong: (a) is inaccurate, (b) involves too much calculation for the accuracy achieved. If the samples are too close, errors due to quantization and numerical model can result [15.4].

## 15.4   Error of application

Some numerical formulae vary from one type of an instrument to another. Take, for example, the estimation of the hole centre coordinates using a roundness instrument and a CMM (coordinate measuring machine) as seen in Figure 15.5.

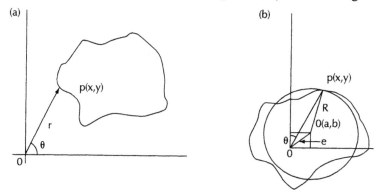

**Figure 15.5**   *Divergence in two metrology procedures: (a) coordinate measuring scheme (b) surface metrology scheme*

At first sight, these should give the same answer but in fact they do not. Neither is wrong. More clearly, the problem is shown in Figures 15.5 (a) and (b). The reason for the discrepancy is that in the roundness machine the origin is within the data (Figure 15.5 (b)), whereas in the CMM the origin is outside the data (see Figure 15.5 (a)). The analysis is as follows.

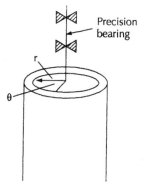

**Figure 15.6**   *Roundness instrument*

The coordinates of the centre $a,b$ are given below from least-squares.

$$a = \frac{2}{N}\sum_{i=1}^{N} r_i \cos\theta_i$$

$$b = \frac{2}{N}\sum_{i=1}^{N} r_i \sin\theta_i$$

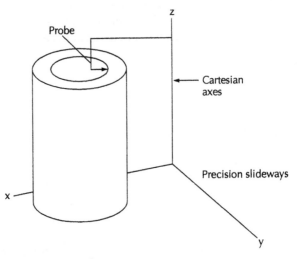

**Figure 15.7**   *Coordinate measuring machine*

The coordinates $a,b$ are twice as large.

$$\left.\begin{aligned} a &= \frac{1}{N}\sum_{i=1}^{N} r_i \cos\theta_i = \frac{1}{N}\sum_{i=1}^{N} x_i \\ b &= \frac{1}{N}\sum_{i=1}^{N} r_i \sin\theta_i = \frac{1}{N}\sum_{i=1}^{N} y_i \end{aligned}\right]$$

Neither is wrong. The different origins were taken into account automatically, but the numerical formulae appear to be wrong!

## 15.5   Representation on drawing

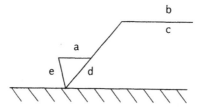

**Figure 15.8**   *Representation on drawing*

Figure 15.8 shows the standard notation where:

a = roughness value or grade;

b = production method;

c = sampling length (cut-off);

d = sampling length of lay;

e = machining allowance (i.e. amount of material that may be removed to the texture specified.

This means that a maximum $R_2$ of 6.3μm and minimum $r$ of 0.2μm is required (the symbol μm or μ is omitted when all the other dimensions on the drawing are either all metric or all British units). Details b to e would be included if relevant or necessary.

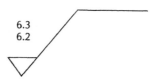

6.3
6.2

**Figure 15.9**   *Example of drawing symbol*

## 15.6   Sample guides

Sample guides are given to help the operator. Table 15.2 shows such a guideline.

**Table 15.2** *Metrological characterization of phase correct filters and transmission bands for use in contact (stylus) instruments*

Gaussian Filter – Short wavelength cut-off $\lambda_s$
      – Long wavelength cut-off $\lambda_c$

| $\lambda$(mm) | $\lambda_s$ ($\mu$m) | $-\lambda_c:\lambda_s$ | Max stylus tip($\mu$m) | Max sampling spacing ($\mu$m) | Min. number of points per cut-off |
|---|---|---|---|---|---|
| 0.08 | 2.5 | 30 | 2 | 0.5 | 160 |
| 0.25 | 2.5 | 100 | 2 | 0.5 | 500 |
| 0.8 | 2.5 | 300 | 2 | 0.5 | 1600 |
| 2.5 | 8 | 300 | 5 | 1.5 | 1666 |
| 8 | 25 | 300 | 10 | 5 | 1600 |

**Table 15.3** *Single values of roughness parameters (ISO 4288)*

Gaussian sampling lengths, for $R_a$, $R_z$ and $R_y$, of periodic profiles.

| $S_m$ mm | | Sampling length mm | Evaluation length Mm |
|---|---|---|---|
| Over | Up to (inclusive) | | |
| (0.013) | 0.04 | 0.08 | 0.4 |
| 0.04 | 0.13 | 0.25 | 1.25 |
| 0.13 | 0.4 | 0.8 | 4.0 |
| 0.4 | 1.3 | 2.5 | 12.6 |
| 1.3 | 4.0 | 8.0 | 40.0 |

## 15.7 Philosophical background to standards in roughness

The following gives the sources of different measurement philosophies within the surface roughness standards (Table 15.3).

There have been two sources of roughness assessment that have been fairly consistent for fifty years. The sources are UK/USA, on the one hand, and Germany/France on the other with Russia in between (Table 15.4).

**Table 15.4**    *Mean line–envelope–motif*

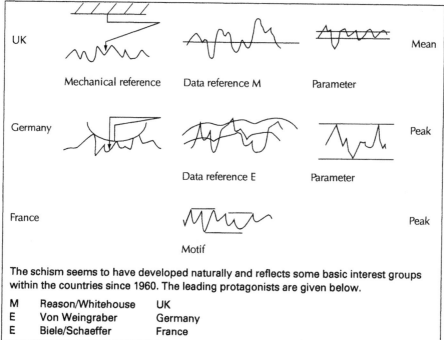

The schism seems to have developed naturally and reflects some basic interest groups within the countries since 1960. The leading protagonists are given below.

| M | Reason/Whitehouse | UK |
| E | Von Weingraber | Germany |
| E | Biele/Schaeffer | France |

Germany emphasized function, whereas UK emphasized manufacture. The former Soviet Union, having the Secretariat, came in between via VS Lukjanov.

In the standards, therefore, there are two distinct threads; one in which the reference line is determined from the peaks with the major parameters being $R_z R + W$ etc. and the other where the reference is determined from all the data and the parameters are averages derived from the data.

The problem has been that there has been no natural rejection of these parameters. Very little positive evidence has emerged or been revealed in a completely unbiased way. Some parameters are in by mistake, some because of usage by some large industry, which is usually automotive.

The note above is given to explain some of the apparent disjointed national and international standards. None of these are necessarily wrong, they represent phases in the evaluation of the subject.

Table 15.5 shows a typical flow chart of the traceability of industrial standards back to the international standards. This shows the procedure for maintaining the standard of calibration. It is relatively easy to set up the calibration but very difficult to keep the standards over time. There has to be a mechanism for monitoring built into the system. This is not just equipment checks but includes personnel,

**Table 15.5**  *Traceability and maintenance of standards*

| | | | |
|---|---|---|---|
| | International | Light Speed Laser | 299792458m/s |
| | NIST  ⎫<br>NIPL  ⎬ → <br>PTB   ⎭ | SINGAPORE ⎫<br>JAPAN<br>AUSTRALIA<br>CHINA  ⎭ ↓ | BIPM<br>FRANCE |
| National<br>Practical<br>Standard<br>Wavelength | NIST NPL<br>interferometer | National Research<br>laboratory. Metrology | Comparison<br>round robin 2 yrs.<br>100mm |
| | Gauge block | Industry/Institute<br>MITUTOYA<br>TOSHIBA<br>JIS limit 0.03μm | |
| | Overseas<br>standard gauge<br>blocks | Industry standard<br>block 3+ yrs | Industry linear<br>scales |
| | | Primary 1 yr | |
| Sample every<br>year<br>Certification | | Secondary 4 × yr | |
| | | Test equipment | |
| | General gauge | Other equipment | Renishaw<br>Heidenheim<br>Industry scales<br>100% |
| | | Total quality<br>procedure | |

procedures, records and so on. In the UK, there is a scheme run from the NPL called NAMAS, which links directly with industry. NAMAS can issue certificates of conformance for measurements taken by NAMAS for customers. Each NAMAS 'centre of excellence' has a director who is responsible for the metrological fidelity. Only named staff are allowed to measure under the auspices of NAMAS. All equipment is registered and periodically checked.

## 15.8 Concluding remarks

This work has highlighted the importance of surface geometry – without detracting from all the aspects of manufacture and function that cover physical and chemical properties of the 'surface skin'. As more processes become 'capable', there will be less need for detailed examination of the surface. However, there is a definite move towards performance parameters. This aspect is now being realized. This opens up a very important aspect of quality that has been rather neglected up to now because of the difficulty of targeting the essential mechanical problems. Emphasis has therefore been upon this area of the subject. The concept of the function map is an attempt to bridge the gap between manufacture and function via surface metrology. Another trend in research will be attempts to predict tribological behaviour in computer simulations of the mating surfaces.

What will be forced into focus will be, and is, the implications of nanotechnology in manufacture.

## References

15.1 Whitehouse D. J. and Reason R. E. *The Equation of the Mean Line of Surface Texture Found by an Electric Wave Filter.* Taylor Hobson Leicester (1965).

15.2 Greenwood J. A. and Williamson J. B. P. *The contact of Nominally Flat Surfaces.* Proc 2$^{nd}$ Int. conf. On Electrical Contacts. Graz, Austria (1964).

15.3 Whitehouse D. J. *The Parameter Rash – Is There a Cure?* Wear (1982).

15.4 Whitehouse D. J. *Handbook of Surface Metrology.* IOP Pub. Bristol (1994).

# Glossary

## Surface Metrology [B5]

**2CR network**  Analogue high pass filter used in early days to block waviness and form signals and thereby allow roughness to be measured.

**A/D converter**  Electronic device to convert an analogue signal to a digital one.

**AA**  Arithmetic average roughness. Used as roughness measure in USA – equivalent to $R_a$ and CLA.

**Abbé error**  Misalignment between the sensitive direction of the metrology instrument and the dimension of the workpiece being measured.

**Abbott–Firestone curve**  Material ratio curve. Sometimes also called bearing ratio curve.

**Abrasive machining**  Machining that uses multiple random grains as the cutting tools.

**Abusive machining**  Machining of such severity as to cause subsurface stresses, usually tensile, and subsurface damage.

**ACF**  Autocorrelation function.

**Adhesion**  Force of attraction at sub-nanometre distances between bodies caused by atomic and molecular forces.

**AFM**  Atomic force microscope. Measures atomic forces.

**Aliasing**  Ambiguity introduced into digital signal due to infrequent sampling, which causes a folding over of the spectrum in the frequency domain.

**Ambiguity function**  A space frequency function based on the Fourier kernel. Achieves, in effect, a two-dimensional correlation.

**Amplitude discrimination**  A way of rejecting small peaks by imposing a minimum height restriction.

**Analytic signal**  A signal that only has positive frequencies. Achieved by use of the Hubert transform.

**Angular distortion**  Distortion of angles in roundness by measuring through apparent centre of part.

**Angular motion**  Angular component of error motion.

| | |
|---|---|
| **Anisotropy** | Degree to which a surface has lay. Deviation from isotropy. |
| **APDF** | Amplitude probability density function. |
| **Areal** | Two-dimensional measurement of area. Sometimes called three-dimensional measurement. |
| **ARMA, MA** | Autoregressive moving average, and moving average. Recursive relationships used in time series analysis. |
| **Aspect ratio** | Ratio of vertical to horizontal magnifications used in instrumentation. |
| **Asperity** | Peak. |
| **Asperity persistence** | The reluctance of asperities to crush under load. Probably due to interaction between asperities. |
| **Assessment length (evaluation length)** | Length of surface over which parameter is assessed. Usually this distance is five sampling lengths. |
| **Auto correlation function** | The expected value of the product of signal with itself displaced. The function is the plot of the expected value for a number of displacements. |
| **Auto covariance function** | As above but with the mean values removed. |
| **Axial magnification** | Line of sight magnification. Movements in the object plane are squared in the image plane. |
| **Axis of rotation** | Actual axis of rotation of spindle of machine tool or measuring device. |
| **Bandwidth** | Effective width of frequency spectrum. Usually measured to half power point. |
| **Barrelling** | Shape of type of cylindrical error. |
| **Basis function** | Unit machining function. |
| **BDRF** | Bidirectional reflectance function. Optical scatter taken over many angles. |
| **Bearing ratio** | Material ratio. |
| **Beat frequency** | Frequency of difference between two signals. |
| **Best fit** | Usually refers to the least-squares criterion. |
| **Beta function** | Function with two arguments used to classify probability density curves. |
| **BIFORE transformation** | Binary Fourier transformation. Related to Walsh function |
| **Bistatic scattering** | Double scattering of ray of light from surfaces. |
| **Born approximation** | Used to determine reflectance coefficients in thin films. |
| **Bosses** | Surface protrusion taken as basis for scatter of light. |
| **Boundary lubrication** | Lubrication regime in which the bodies are separated by a molecularly thin film that reduces friction but down or support load. |
| **Box function** | A function representing a running average of data. |
| **Bragg equations** | Equations relating to the scattering of X-rays from a crystal. Elastic scattering. |
| **Bragg scattering** | X-ray scattering from crystal lattice |
| **Brillouin scattering** | Non-elastic scattering – acoustic mode of vibration. |
| **Brownian movement** | Random movement similar to random movement of molecules. |
| **BUE** | Built-up edge of material left on tool during cutting. |
| **Burnish** | Plastic movement of material by cutting tool, which usually smooths topography. Often produced in diamond turning. |
| **CAD** | Computer aided design. |

| | |
|---|---|
| **Calibration chain** | The chain of calibration from international to workshop standards. |
| **Calibration specimens** | Specimens used to calibrate roughness measuring instruments, usually for $R_q$. |
| **Calliper** | The basic system for a metrology instrument. |
| **Capability** | The guaranteed performance achievable by a machine tool. |
| **Cartesian coordinates** | Coordinate plotted on orthogonal axes. |
| **Caustics** | Reflections off surfaces caused by local curvatures. |
| **CBN** | Cubic boron nitride – cutting tool material. |
| **Cepstrum** | Inverse Fourier transform of the logarithm of the power spectrum. |
| **Characteristic depth** | Depth of profile from motif. |
| **Characteristic function** | The Fourier transform of the logarithm of the power spectrum. |
| **Chatter** | Vibration imparted to a workpiece by elements of the machine tool being too flexible. |
| **Chebychev polynomial** | Type of polynomial that fully uses the separation between two boundaries. |
| **Chirp signal** | A signal whose frequency changes as the square of time. |
| **Chordal** | Measurement of roundness taken off chords of workpiece. |
| **Circularity** | Roundness. |
| **CLA** | Centre line average. Old roughness parameter equivalent to $R_a$. |
| **CMM** | Coordinate measuring machine. |
| **Coherence discrimination** | System that allows interference to be detected. |
| **Coherence modulation** | This is the form given to the envelope of a fringe pattern due to the finite bandwidth of the source. |
| **Coherence spatial** | Degree to which two rays from different spatial positions of the source are in phase. |
| **Coherence temporal** | Degree to which a signal is monochromatic. |
| **Compliance** | Elastic deformation of object when under load. |
| **Compression ratio** | Difference in magnification between horizontal and vertical axes. |
| **Compton scattering** | Non-elastic scattering. |
| **Concentricity** | Two x eccentricity. Locus of the centre of selected figure rotating around selected datum e.g. axis. |
| **Condition function** | Boundary conditions in exchange algorithms. |
| **Confidence interval** | Boundaries on probability of a value to within a given significance. Any geometric figure governed by second degree equations. |
| **Conicity** | The departure of an object from a true cone. |
| **Conversion efficiency** | The efficiency of a transducer in converting energy from one form into another. |
| **Convolution** | A type of integration of two functions in which the variable of integration gets folded. |
| **Correlation** | The expected value of the product of two variables normalized with respect to their standard deviations. |
| **Correlation length** | The lag value over which the autocorrelation function falls to a small value, usually 10% or 1/e. |
| **Creep** | Difference in velocity between different points within contact region of rolling ball. |
| **Crest** | Taken to mean peak. |

| | |
|---|---|
| **Critical angle** | The angle below which internal reflection takes place. |
| **Critical distance in ductile grinding** | The depth of cut that allows the mode of grinding to be plastic and not fractural. |
| **Curvature** | Reciprocal of the radius of curvature. Sometimes approximated as the second differential. |
| **Cut-off length** | The wavelength along the surface that corresponds to the sampling length. |
| **Cylindricity** | Departure of workpiece from a truly cylindrical shape. |
| **D ratio** | Ratio of surface roughness to film thickness in pitting. |
| **D sight** | Whole sight techniques incorporating reflection from a diffuser, used to show up flaws. |
| **DAF** | Discrete ambiguity function. |
| **Damping factor** | Term representing energy loss in a second-order differential equation. |
| **Defect** | A departure from the expected statistics of a surface. |
| **Degree of freedom** | An independent movement. |
| **Designer surface** | A surface made specifically for a given function. |
| **DFT** | Discrete Fourier transform. |
| **DFTC** | Departure from true circle. |
| **Diagonal sampling and analysis** | Three-point digital analysis. |
| **Diametral** | Variations across diameters in roundness measurement. |
| **Difference operators** | Difference between discrete measurements. Can be central, forward, or backward. Used in numerical analysis. |
| **Differential logarithm errors** | The logarithms of the errors in a propagating formula for errors. |
| **Differential spatial damping coefficient** | The damping coefficient expressed in spatial terms. |
| **Diffuse reflection** | Light scattered from a surface at angles other than the specular angle. |
| **Digitization** | Taking discrete values of an analogue waveform usually at equal intervals. |
| **Dimensional metrology** | The measurement of the linear dimensions of workpieces. Usually including angles. |
| **Dirac comb** | A train of impulses. Can be used to represent sampling discretely. |
| **Directionality** | A measure of the way in which crests are pointing away from the normal. |
| **Discrete parameter** | A parameter of a surface obtained from the digital values. |
| **Discrete properties** | Properties of the digital form of surface. |
| **DMT** | Model based on Derjaguin to explain adhesion between objects. |
| **Dual** | Alternative to primal method in exchange mechanisms. |
| **Ductile grinding** | Grinding in which the mode of material removal is plastic. |
| **Dynamic interaction** | Contact of two bodies, usually under lateral movement. |
| **E system** | System of measurement based on rolling ball across surface and measuring from the locus of the lowest point. |
| **Eccentricity** | The distance between the centre of workpiece and centre of rotation of measuring instrument. |
| **ECM** | Electrochemical machining. |

| | |
|---|---|
| **EDM** | Electrodischarge machining. |
| **EHD** | Elastohydro dynamic lubrication. |
| **EHL** | Elastohydro dynamic lubrication. |
| **Eigen vector** | Made of solution of a matrix. |
| **Eigen vector elastic scattering** | Scattering of radiation from an object that does not result in a change of wavelength. |
| **Ellipticity** | The maximum ratio of two orthogonal axes describing a near circular part. |
| **Energy gradient in instruments** | The rate at which energy is transferred as a function of displacement. |
| **Engineering metrology** | The overall subject of measurement as viewed from an engineering point of view. |
| **Envelope** | Curve generated according to a set of rules tangentially connecting peaks and /or valleys. Used as reference from which to measure roughness or waviness. |
| **Envelope methods** | Methods based on envelopes. |
| **Epitrochoid** | Shape of the stator of the Wankel engine. |
| **Equal weight techniques** | Digital samples based on equal area rather than equal intervals. |
| **Ergodic** | Statistical situation where a temporal average is equal to a spatial average. |
| **Error separation methods** | A technique in which the errors of a specimen and the instruments reference movement are isolated simultaneously. |
| **Errors** | Deviation from intended shape or size. |
| **Errors of form** | Long wavelength. Geometric deviation from the intended geometric shape. |
| **Expectation** | Average value statistical expectation. |
| **Extrapolation** | Technique to estimate value of a function outside its given range. |
| **F number** | Means of describing lens performance. Focal length divided by lens aperture. |
| **$F$ test** | Fisher test for variance. |
| **Face motion** | Component of error motion of machine tool orthogonal to the axis of rotation. |
| **Factorial design** | Specialist form of experimentation designed to isolate the effect of the independent variables. |
| **FECO fringes** | Fringes of equal chromatic order. |
| **FFT** | Fast Fourier transform. |
| **Fibre optic** | Thin strand of optical transparent material along which information or electromagnetic energy can be transmitted. |
| **Fidelity** | The closeness of a measured signal to the original signal. |
| **Filter cut-off** | Frequency (or wavelength) corresponding to 50% attenuation. In the past this has been 75%. |
| **Fingerprint in manufacture** | A set of surface parameters completely defining the process and machine tool. |
| **Finish machining** | Final machining to achieve desired surface texture and dimension. |
| **Flash temperature** | Temperature of contacting asperities when scuffing occurs. |
| **Flatness** | Departure of workpiece from true flatness. |
| **Flaw** | Deviation from the expected statistics of the workpiece. |

| | |
|---|---|
| **Flicker noise** | Electrical noise inversely proportional to frequency. |
| **Flying spot microscope** | A microscope in which a small source of light and a small detector localized to it are synchronized. |
| **Follower** | A measuring system in which the surface geometry is followed by the measuring instrument. |
| **Force loop** | The path along which forces act in a metrology instrument or machine tool. Usually drive forces and inertial forces. |
| **Fourier transform** | Transformation depicting summation of sine waves. |
| **Fractal dimension** | The parameter depicting the scale of size. Given by 1.2(5-n) where n is power law of spectrum. |
| **Fractal property** | A multiscale property in which the parameters are the same for each scale of size. |
| **FRASTA** | Fracture surface topography analysis. Use of surface topography to assess fracture mechanics. |
| **Fraunhoffer diffraction** | Intensity pattern in the back focal plane of a lens. Usually the image of the source of light modulated by surface scatter. Typified by superposition of plane wavefront. |
| **Fresnel diffraction** | Intensity pattern produced by point source. Typified by spherical wavefronts. |
| **Fretting** | Wear produced by contact of bodies with lateral movement and vibration usually involving trapped debris. |
| **Function** | The application of a workpiece. |
| **Functional parameter** | Parameter that is important in a functional context. |
| **Fundamental motion** | Error motion of machine tool. |
| **Gabor transforms** | Space frequency function based on Gaussian weighting function. The factorial distribution. |
| **Gamma distribution** | The factorial distribution. |
| **Gaussian** | Distribution having an exponential with squared terms of the variable. |
| **Gaussian filter** | Filter having a Gaussian weighting function or frequency transmission characteristics. |
| **Geometric surface** | The ideal surface defined by the drawing or specification. |
| **Glossmeter** | Instrument for measuring light scatter from surface. |
| **Goodness of fit test** | Chi-squared test of the equivalence of hypotheses. |
| **Gram-Charlier series** | Method of characterizing a probability density function. |
| **Grubler's equation** | Mobility of linkages in terms of links and joints. |
| **alphaGuard band** | Attenuation band in frequency characteristics isolating two blocks. |
| **Hadamard function** | Clipped signal technique. |
| **Harmonic weighting function** | Function describing the distortion of measured harmonic coefficients by a measuring technique. |
| **Hartley transform** | Transform similar to Fourier transform, only having no complex components. |
| **Hatchet stylus** | Stylus used in measuring form. The hatchet shape integrates out the roughness. |
| **Hazard** | Rate of change of failure. |
| **Helical track** | Traverse technique sometimes used to measure cylindricity. Combines one linear and one rotary motion. |
| **Helmhotz-Kirchoff integral** | Equation that describes electromagnetic radiation properties. |

| | |
|---|---|
| **Hermite polynomials** | Related to differentials of Gaussian distribution. |
| **Heterodyne methods** | Techniques using two independent modes of measurement such as two frequencies or two polarizations. |
| **Hexagonal sampling** | Sampling pattern using seven points. |
| **High pass filter** | Filter technique passing only small wavelengths. |
| **Hill climbing** | Technique of optimization. |
| **Hip prosthesis** | Replacement hip joint of partial spherical shape. |
| **Hole** | A closed contour on an areal map. |
| **Holography** | Photograph containing phase information trapped by means of reference beams. |
| **Homodyne methods** | Technique using similar modes of measurement. |
| **Hybrid parameter** | A parameter such as slope derived from two or more independent variables. |
| **Hypergeometric function** | A three argument function used for characterization. |
| **Hypotrochoid** | A geometric figure of the family of trochoids. |
| **Impulse response** | The response of a system to an impulse. |
| **Inferior envelope** | Line through a succession of valleys. |
| **Instantaneous axis** | Centre of rotation at a given time. |
| **Instrument capability** | Guaranteed performance specification of instrument. |
| **Integrated damping** | Damping coefficient optimized over a wide frequency band. |
| **Interpolation** | Derivation of the value of a signal between two known points. |
| **Intrinsic equation** | Equation developed from the profile itself. |
| **Inverse scatter problem** | The deduction of the surface statistics from the statistics of the scattered wavefront. |
| **Isotropy** | The uniformity of pattern of the surface. |
| **Iterative relation** | A relationship between values of system output and system input for various instances in time or space. |
| **Jacobian** | Relationship between the differential coefficients in different domains. |
| **JKR** | Model based on Johnson to explain adhesion between solids. |
| **Johnson noise** | Electronic noise dependent on input resistance. |
| **Joint probability density function** | Function describing statistical behaviour at an infinitesimal point. When integrated between two points gives probability. |
| **JPDF** | As above. |
| **$K$ value** | Semi-empirical number found by Jakeman for fractal type surfaces. |
| **Kelvin clamp** | Method of achieving six constraints by means of kinematic location. |
| **Kinematics** | Laws of constraint of a free body. |
| **Kirchoff laws** | Laws of optical behaviour. |
| **Kitagawa plot** | Logarithmic plots of variables. |
| **Kurtosis** | Fourth central moment of a distribution. |
| **Lagrangian multipliers** | Method used to evaluate best fit coefficients, usually with conditional differentials. |
| **Langmuir-Bloggett films** | Molecularly thin films. |
| **Laplace transform** | Transform used to determine the response of a system to generalized and impulsive inputs. |
| **Laser waist** | Minimum focused width of laser beam. |
| **Lateral roughness** | Roughness at right angles to movement. |

| | |
|---|---|
| Lay | Pattern of areal surface finish. Direction of the prevailing texture of the surface, usually determined by the method of manufacture. |
| least-squares centre | Centre position determined by a least-squares method. |
| least-squares cylinder | A cylinder derived from the least-squares straight line determined from the centres of the least-squares circle of each measured profile. |
| Length of profile | Actual length of profile expressed in terms of differentials. Levelling depth $R_p$ maximum peak value in sampling length. |
| Likelihood | Expected value. |
| Limacon | True equation of eccentric circular part when eccentric to rotate in axis of roundness instrument. |
| Linear phase filter | Filter having an impulse response with an axis of symmetry about a vertical axis. |
| Linear programming | Technique in which variables are related linearly. |
| Lobing | Undulations on a nominally round workpiece, produced usually by clamping or chatter. |
| Locus | Line produced by some procedure. |
| Log normal function | Logarithm of the Gaussian distribution. Usually refers to distribution of extrema. |
| Long crestedness | Measurement of areal bandwidth of surface according to Longuett-Higgins. |
| Longitudinal profile | Profile resulting from the intersection of a surface by a plane parallel to the lay. |
| Low pass filter | Filter passing only long wavelengths. |
| LSC | Least-squares circle. |
| LVDT | Linear voltage differential transformer. |
| M system | Measurement system based on mean line references. |
| Map | Areal coverage of surface. |
| Markov process | Sequence in which current value depends only on the previous one. |
| Material ratio | Ratio of material to air at any given level relative to mean line. |
| Maximum material condition | Basis of tolerancing system. Minimizes the amount of machining. |
| MC | Minimum circumscribed circle. |
| MCE | Mean crest excursion. |
| Mean line of roughness profile (M) | Reference line in the evaluation length such that the area enclosed between it and the profile has equal positive and negative values. |
| Mechanical loop | Loop linking the mechanical elements, usually of the metrology calliper. |
| Meter cut-off | Same as cut-off, filter cut-off. |
| Method of exact fractions | Method using multiple wavelengths in which, by measuring fracture of fringes, step heights can be measured. |
| Metrology loop (measuring loop) | The linking of all the mechanical elements making up the metrology calliper. |
| MFM | Magnetic force microscope. |
| MI | Minimum inscribed circle. |
| Midpoint locus | Locus produced by plotting outcome of running average procedure. |

| | |
|---|---|
| **Minimum phase** | System in which amplitude and phase are explicitly related. |
| **Minimum zone centre** | Centre position based on minimum zone procedure. |
| **Minimum zone references** | Zonal methods that have minimum separation. |
| **MND** | Multinormal distribution or multi-Gaussian distribution. |
| **Moiré fringes** | Fringes produced by overlaying of true structures of equal spacing at an angle. |
| **Motif** | Procedure for producing envelopes for use as reference lines. |
| **Motion copying** | Machining in which tool is forced to follow a predetermined geometric path. |
| **Multiprocess** | A manufacturing process such as plateau honing in which more than one process makes up the final surface. |
| **Multivariate normal distribution** | Same as MND |
| **MZ** | Minimum zone circle. |
| **Non-Newtonian** | System in lubrication in which viscosity changes with pressure. |
| **Normal equations** | Differential equations produced from the summation of the squared deviations of observed and ideal values. |
| **Numerical analysis** | Analysis of the discrete form of the surface by numerical rules. |
| **Numerical aperture** | Factor influencing resolution and depth of focus of optical element. |
| **Numerical model** | Discrete model of surface feature, i.e. three-point model of peak. |
| **Nyquist criterion** | Frequency criterion advocating sampling at twice the highest frequency of interest. |
| **Objective speckle** | Intensity pattern produced when rough surface is illuminated by a laser. |
| **One number parameter** | The concept of attempting to judge the complete merit of a waste piece by one number. |
| **OOR** | Out-of-roundness. |
| **Optimized datum plane** | Reference plane obtained by optimized procedure. |
| **Ordinate** | Height measurement, usually in discrete form. |
| **Out-of-roundness** | OOR |
| **Ovality** | Maximum difference between length of two axes through object centre, usually but not necessarily at right angles. |
| **Parallax** | Produces distortion in scanning electron microscopes. |
| **Parameter** | Feature to be quantified. |
| **Parameterization** | The process of representing a function in terms of an indirect variable. |
| **Parasitic movement** | Spurious secondary movement or vibration. |
| **Pareto curve** | Most effort should be expended in most significant effects. Pareto's law. |
| **Partial arc** | An incomplete curve, usually circular. |
| **PCF** | Pressure concentration factor. |
| **Peak** | Maximum of profile between two adjacent valleys. |
| **Peak density** | Number of peaks per unit distance. |
| **Peak to valley (roundness)** | Radial separation of two concentric circles, which are themselves concentric to the reference and totally enclose the measured profile. |
| **Periodic profile** | Profile that can be described by a periodic function, e.g. turned, milled. |

| | |
|---|---|
| **Perturbation methods** | Method for solving complex differential equations using small variation. |
| **Phase corrected filter** | Filter with no phase distortion. |
| **PHL** | Plastohydrodynamic lubrication. |
| **Physical metrology** | Metrology of physical parameters such as hardness and stress. |
| **Pick-up** | Device used to detect information. |
| **Planimeter** | Analogue device for measuring area. |
| **Plasticity index** | Numerical index for predicting plastic flow. |
| **Plug gauge centre** | Centre found by maximum inscribing circle algorithm. |
| **Poincaré section** | Two-dimensional section in 3D state space. |
| **Polar co-ordinates** | Use of radial variations as a function of angle. |
| **Polar distortion** | Deviations in shape of circle produced as a result of nature of roundness instruments. |
| **Polygonation** | The breakdown of a circular signal into a polygon form by means of triangles. |
| **Polynomial fit** | The fitting of a smooth polynomial through asset of discrete ordinals. |
| **Power spectral density** | The limiting value of the period gram – the square of the Fourier coefficients. |
| **Pressure copying** | Topography generated as a result of pressure input. |
| **Pressure distribution** | Variation of pressure as a function of distance within the contact zone. |
| **Prestressing** | Application of fixed stress usually to position working zone in a more predictable region. |
| **Primal** | Conventional method of tackling exchange algorithms. |
| **Primary cutting edge** | Cutting along the track of cutting tool. |
| **Primary profile** | The profile resulting after the real profile has been investigated by a finite stylus. |
| **Process parameters** | Parameters such as depth of cut and feed typical of a manufacturing process. |
| **Profile** | A section taken through a workpiece. |
| **Profile length** | The true length of the profile as opposed to its $x$ dimension. |
| **Profile parameter** | An attempt to quantify the profile according to some procedure. |
| **Propagations of errors** | The resultant error in a system in terms of errors in its constituents. |
| **PSD** | Power spectral density. |
| **Pseudo-kinematics** | Kinematic design that allows a certain amount of elastic averaging. |
| **Pseudo-random sequence** | A random sequence generator that repeats after a limited number of operations. |
| **Quantization** | The breaking down of an analogue height into discrete binary units. |
| **Quill** | Housing for roundness instrument spindle. |
| **Radial** | Coordinate measured from a point irrespective of direction. |
| **Radial motion** | Movement of tool or workpiece in radial direction in error motion. |
| **Rain fall count** | Method of adding up effective count in fatigue. |
| **Raised cosine** | Unity added to cosine to make value always positive. Used in lag windows. |

| | |
|---|---|
| **Rake** | Angle of tool presented to workpiece. |
| **Ramon scattering** | Non-elastic scattering – optical mode of vibration. |
| **Random error** | Error that cannot be predicted. Sometimes called stochastic error. |
| **Random process analysis** | Autocorrelation, power spectral density and probability density. |
| **Random profile** | Profile that can be described by a random function, e.g. ground, shot blasted. |
| **Range (measuring range)** | Usually taken to be the range over which a signal can be obtained as some function, not necessarily linear, of displacement. |
| **Rautiefe** | Original name for $R_t$. |
| **Ray tracing** | Optical behaviour based on geometrical optics. |
| **Rayleigh criterion** | Criterion of optical resolution. |
| **Reaction** | Force from surface responding to stylus force. |
| **Readability** | The ability to detect a change in value of a variable and represent it meaningfully. |
| **Real surface** | Surface limiting the body and separating it from the surrounding media. |
| **Real surface profile** | The line of intersection of a real surface and a plane. |
| **Recursive filters** | Filter whose current output can be expressed in terms of current inputs and past outputs. |
| **Reference circle** | Perfect circle of same size as workpiece from which deviations of roundness are measured. |
| **Reference cylinder** | Perfect cylinder of same size as workpiece from which deviations of cylindricity are measured. |
| **Reference line** | Line constructed from geometrical data from which certain features can be measured. Most often used to separate roughness and waviness. |
| **Reference line** | The line relative to which assessment of profile parameters are determined. |
| **Reference surface** | The surface relative to which roughness parameters are determined. |
| **Referred part** | A perfect part estimated from existing data. |
| **Regression** | The finding of, usually linear, relationships between variables by least-squares methods. |
| **Relaxation** | The tendency of a body to sink to a position of lower potential energy with time. |
| **Relocation profilometry** | Techniques used to ensure that a workpiece can be positioned time after time in the same place. |
| **Residual stress** | Stresses usually left in the subsurface of a workpiece as a result of machining. |
| **Resolution** | The ability to separate out two adjacent lateral details. |
| **Reversal methods** | Method used to separate out reference and workpiece values of geometry by changing the dependence of one on another. |
| **Reynold's equation** | Equation governing the pressure between two bodies in relative lateral motion separated by a fluid film. |
| **Reynold's roughness** | Roughness that is small compared with the film thickness. |
| **Ring gauge centre** | Centre found by means of the minimum circumscribing circle algorithm. |

| | |
|---|---|
| **RMSCE** | Root mean square crest excursion. |
| **Roughness** | Marks left on surface as a result of machining. |
| **Roughness evaluation length (1)** | Length of the reference line over which the mean values of the parameters are determined. |
| **Roughness profile** | A real profile after modification by filtering, e.g. elec. filter, stylus. |
| **Roundness typology** | Characterization of a workpiece according to the coefficients of the Fourier components of the roundness signal. |
| **Running-in** | The process of easing a workpiece into a steady state wear regime. |
| **Runout** | Total deviation as seen by instrument probe. |
| *S* **number** | Sputtering rate. |
| **Sampling** | The process of taking discrete values of an analogue signal. |
| **Sampling length (1) (roughness)** | The basic length over which roughness is measured and is equal to the cut-off length. |
| **Sampling rate** | The rate at which discrete measurements are made. |
| **Scalar theory** | Theory of electromagnetic radiation in which phase and amplitude of the wavefront are important. |
| **Scatterometer** | Instrument for measuring the light scattered from a surface. |
| **SCM** | Scanning confocal microscope. |
| **Scuffingwear`** | Wear produced by failure due to thermal runaway. |
| **Secondary cutting edge** | Cutting produced in the axial direction. |
| **Seidal aberration** | Errors in optical wavefront produced by second order faults in the optical system. |
| **Self-affine** | Property that allows equivalence if one axis is changed. |
| **Self-similarity** | Property that allows equivalence of properties at all scales of size. |
| **SEM** | Scanning electron microscope. |
| **Sensitive direction** | Direction in which the rate of change of energy conversion is a maximum. |
| **Sensitivity** | The energy or information conversion rate with displacement. |
| **Shadowing** | Produced by detail on the surface being washed out by oblique angle illumination. |
| **Sheppard's correction** | Numerical noise produced as a result of finite quantization. |
| **Shot noise** | Noise produced by variation in standing electric currents akin to Brownian motion. |
| **Simplex method** | An exchange algorithm used in linear programming. |
| **Simpson's rule** | Rule for numerical integration based on fitting parabolas between points. |
| **Singular value decomposition** | Method of rearranging matrices so as to be easy to solve. |
| **Skew** | Third central moment of a distribution. |
| **Skew limacon** | Limacon shape found in cross section of tilted cylinder. |
| **Skid** | Mechanical filter, used as reference. |
| **Space-frequency functions** | Functions such as Wigner that have two arguments, one in space and one in frequency. |
| **Spalling** | Another word for pitting. |
| **Spark-out** | A means of ensuring that all possible elements on a grinding wheel actually cut. |

| | |
|---|---|
| **Spatial wavelength** | Wavelength on surface expressed in millimetres, not seconds. |
| **Spazipfel** | Triangular piece of material left on surface during cutting. |
| **Speckle** | Grainy appearance of object produced in laser light by aperture-limited optic. |
| **Spectroscopy** | The breaking down of a signal into frequency components. |
| **Specular reflection** | Light scattered at the reflected angle. |
| **Speed ratio** | Ratio of workpiece circumference speed to that of wheel circumference speed. |
| **Sphericity** | Departure of geometry from a true sphere. |
| **Spline function** | Mathematical function usually of cubic nature derived from use of elastic wooden spline. |
| **Squareness** | Errors in position of two planes nominally at right angles. |
| **Squeeze film** | Lubrication film in the presence of normal vibration. |
| **Straightness** | Departure of surface geometry from a straight line. |
| **Stribeck diagram** | Shows the degree of lubrication. |
| **Standard cut-off (wavelength)** | Nominally equal to 0.8 mm or 0.03". |
| **Standard sampling length** | As above. |
| **Standardization** | Limitation of procedures to those agreed. |
| **State-space** | Velocity displacement display of system output. |
| **Static interaction** | Two body contact without lateral movement. |
| **Stiefel exchange mechanism** | Mechanism used in linear programming. |
| **Stieltjes correlator** | Correlator using one channel of clipped signals working on reduced bandwidth. |
| **Stiffness`** | Slope of force displacement curve of contact. |
| **STM** | Scanning tunnelling microscope. |
| **Stoke's roughness** | Roughness that is a significant proportion of film thickness. |
| **Strange attractor** | Word given to apparent focus in space state diagram. |
| **Stratified process** | Finishing produced by more than one process. Processes usually present at different heights in the topography. |
| **Structure function** | Expected values of square of difference between two values of waveform. |
| **Stylus** | Mechanical implement used to contact workpiece and to communicate displacement to transducer. |
| **Stylus tree** | Rod containing a number of styli at different axial positions. |
| **Subjective speckle** | Speckle effect produced when surface illuminated by a laser is imaged. Speckle is in image plane. |
| **Subsurface characteristics** | Physical characteristics immediately below geometric boundary of surface. |
| **Summit** | Areal peak. |
| **Superior envelope** | Line connecting suitable peaks. |
| **Superposition integral** | Linear summation of properties. |
| **Suppressed radius** | Workpiece size removed in order to enable roundness instrument to measure the surface skin with high accuracy. |
| **Surface integrity** | Properties taken sometimes to mean physical and geometrical properties in the US. Usually stress related properties only. |
| **Surface roughness** | Irregularities in the surface texture that are inherent in the manufacturing process but excluding waviness and errors of form. |

| | |
|---|---|
| **Surface texture** | The geometry imparted. |
| **Symbolic notation of errors** | Symbolic method of writing error propagation equations. |
| **System response** | Output of the system to a given point. |
| **Systematic error** | Error that is attributable of deterministic cause and hence can be predicted. |
| **Tactile response** | Instrument that uses contact as a means of collecting information. |
| **Taper** | Cylindricity in which the radial value varies with height linearly. |
| **TEM** | Transmission electron microscope. |
| **Tetragonal sampling** | Sampling with five-point analysis. |
| **Tetrahedron ball frame** | Frame built in form of tetrahedron to calibrate coordinate. |
| **Theoretical surface finish** | Surface finish determined totally from process parameters such as feed, depth of cut, tool radius. |
| **TIS** | Total integrated scatter. |
| **Tolerance** | Range of allowable dimensional or other parameter variation of workpiece. |
| **Top down mechanics** | Mechanisms simulating contact. |
| **Topography** | The study of surface features. |
| **Topothesy** | Length of chord on surface having average slope of one radian. |
| **Traceability** | The linking of measurement procedures to national standards. |
| **Trackability** | The ability of stylus system to follow geometry without losing contact. |
| **Transducer** | Energy conversion system. |
| **Transfer function** | Dynamic relationship between input and output of system. |
| **Transmission characteristics** | Plot of transfer function. |
| **Transverse profile** | Profile resulting from the intersection of a surface by a plane normal to the surface lay. |
| **Trapezoidal rule** | Method of numerical integrating linking adjacent points linearly. |
| **Tribology** | Science of rubbing parts. |
| **Trigonal sampling** | Sampling pattern using four-point analysis of height in a profile. |
| **Truncation** | The curtailing of aspects of a function. In particular, the cutting off of height in a profile. |
| **Typology** | A characterization. |
| **Union Jack pattern** | A pattern of measurements used to cover surface plates in an economic and self-regulating way. |
| **Unit event of machining** | Basic unit relating finish to average grain thickness or cutting particle size. |
| **UPL, LPL, LVL, UVL** | Upper peak limit etc. |
| **Valley** | Minimum of the profile. |
| **Van Der Waal's forces** | Atomic forces. |
| **Vector theory** | Electromagnetic theory incorporating polarization. |
| **Vee block method** | Method of calibrating surface texture instruments. |
| **Vernier fringes** | Fringes produced by overlay of two scales of slightly different spacing. |
| **Vibrating table method** | Method of calibrating surface texture instruments. |
| **Walsh function** | Transform using square waves rather than sine waves. |
| **Wankel engine** | Rotary engine with stator shaped as epitrochoid. |

| | |
|---|---|
| **Wavelet transform** | Space frequency function. |
| **Waviness** | Geometric features imparted to workpiece usually by errors in the machine tool motion. |
| **Weibull distribution** | Graph of hazard with time. |
| **Weierstarss function** | Fractal function. |
| **Weighting function** | Set of weighting factors. |
| **Wein's law** | Law relating maximum radiation to temperature. |
| **Whole field technique** | Method of showing up distortion. |
| **Wigner distribution function** | Energy required to release electrons from surfaces. |
| **Work function** | Energy required to release electrons from surfaces. |
| **Wringing** | Sticking of smooth surfaces in contact. Presence of fluid film necessary. |
| *z* **transform** | Discrete transformation technique for dealing with the dynamics of discrete systems. |
| **Zeeman splitting** | Splitting of laser frequency by magnetic field. |
| **Zernicke polynomials** | Polynomials used in specifying optical performance. |
| **Zero crossing** | Crossing of the mean line. |

# Index